An outstanding volume of non-Eurocentric historical-sociological essays that advances an extremely powerful critique of the English School's commitment to what I call the 'Eurocentric big bang theory of world politics' – that the big bang of modernity exploded autonomously within Europe and that thereafter European 'civilization' expanded outwards in a non-problematic way to remake the world in its own image – by deploying the antidote of bringing Eastern agency back into the story of the long-term development of world politics.

John M. Hobson, Professor of Politics and International Relations, University of Sheffield, UK

Perhaps '1648 and all that' fundamentally distorts our understanding of the history of international relations. By rediscovering early modern and non-western international orders, this expert team sets out a challenge to IR scholarship. It is especially welcome and important in making sense of the profound changes currently underway.

Ian Clark, E. H. Carr Professor of International Politics, Aberystwyth University

Maps centered on the South Pole are amusing, but international relations books centered on the non-Western world are essential. From its origins, international relations theory has been a Western enterprise with other parts of the world largely ignored or analyzed through parochial and often inappropriate conceptions. Suzuki, Zhang, Quirk and their collaborators turn the tables and offer perspectives on Europe from East Asia, the Middle East, Africa and Latin America, and generally at a time when European interlopers were unable to impose their preferences on these cultures. They effectively demonstrate the importance of non-Western cultures and their ideas in shaping global history

Richard Ned Lebow, Professor of International Political Theory, King's College London, UK

International Orders in the Early Modern World

This book examines the historical interactions of the West and the non-Western world, and investigates whether or not the exclusive adoption of Western-oriented 'international norms' is the prerequisite for the construction of stable and enduring international orders.

This book sets out to challenge the Eurocentric foundations of modern International Relations scholarship by examining international relations in the early modern era, when European primacy had yet to develop in many parts of the globe. Through a series of regional case studies on East Asia, the Middle East, Africa, Latin America, and Russia written by leading specialists of their field, this book explores various historical patterns of cross-cultural exchange and civilizational encounters. The chapters of this book document and analyse a series of regional international orders that were primarily defined by local interests, agendas and institutions, with European interlopers often playing a secondary role. These perspectives emphasize the central role of non-European agency in shaping global history, and stand in stark contrast to conventional narratives revolving around the 'Rise of the West', which tend to be based upon a stylized contrast between a dynamic 'West' and a passive and static 'East'.

Focusing on a crucial period of global history that has been neglected in the field of International Relations, *International Orders in the Early Modern World* will be of interest to students and scholars of international relations, international relations theory, international history, early modern history and sociology.

Shogo Suzuki is Senior Lecturer in the Department of Politics at the University of Manchester, UK.

Yongjin Zhang is Professor of International Politics at the University of Bristol, UK.

Joel Quirk is Senior Lecturer in the Department of Political Studies at the University of the Witwatersrand, South Africa.

The New International Relations

Edited by Richard Little, *University of Bristol*, Iver B. Neumann, *Norwegian Institute of International Affairs (NUPI), Norway* and Jutta Weldes, *University of Bristol.*

The field of international relations has changed dramatically in recent years. This new series will cover the major issues that have emerged and reflect the latest academic thinking in this particular dynamic area.

International Law, Rights and Politics
Developments in Eastern Europe and the CIS
Rein Mullerson

The Logic of Internationalism
Coercion and accommodation
Kjell Goldmann

Russia and the Idea of Europe
A study in identity and international relations
Iver B. Neumann

The Future of International Relations
Masters in the making?
Edited by Iver B. Neumann and Ole Wæver

Constructing the World Polity
Essays on international institutionalization
John Gerard Ruggie

Realism in International Relations and International Political Economy
The continuing story of a death foretold
Stefano Guzzini

International Relations, Political Theory and the Problem of Order
Beyond international relations theory?
N.J. Rengger

War, Peace and World Orders in European History
Edited by Anja V. Hartmann and Beatrice Heuser

European Integration and National Identity
The challenge of the Nordic states
Edited by Lene Hansen and Ole Wæver

Shadow Globalization, Ethnic Conflicts and New Wars
A political economy of intra-state war
Dietrich Jung

Contemporary Security Analysis and Copenhagen Peace Research
Edited by Stefano Guzzini and Dietrich Jung

Observing International Relations
Niklas Luhmann and world politics
Edited by Mathias Albert and Lena Hilkermeier

Does China Matter? A Reassessment
Essays in memory of Gerald Segal
Edited by Barry Buzan and Rosemary Foot

European Approaches to International Relations Theory
A house with many mansions
Jörg Friedrichs

The Post-Cold War International System
Strategies, institutions and reflexivity
Ewan Harrison

States of Political Discourse
Words, regimes, seditions
Costas M. Constantinou

The Politics of Regional Identity
Meddling with the Mediterranean
Michelle Pace

The Power of International Theory
Reforging the link to foreign policy-making through scientific enquiry
Fred Chernoff

Africa and the North
Between globalization and marginalization
Edited by Ulf Engel and Gorm Rye Olsen

Communitarian International Relations
The epistemic foundations of international relations
Emanuel Adler

Human Rights and World Trade
Hunger in international society
Ana Gonzalez-Pelaez

Liberalism and War
The victors and the vanquished
Andrew Williams

Constructivism and International Relations
Alexander Wendt and his critics
Edited by Stefano Guzzini and Anna Leander

Security as Practice
Discourse analysis and the Bosnian War
Lene Hansen

The Politics of Insecurity
Fear, migration and asylum in the EU
Jef Huysmans

State Sovereignty and Intervention
A discourse analysis of interventionary and non-interventionary practices in Kosovo and Algeria
Helle Malmvig

Culture and Security
Symbolic power and the politics of international security
Michael Williams

Hegemony & History
Adam Watson

Territorial Conflicts in World Society
Modern systems theory, international relations and conflict studies
Edited by Stephan Stetter

Ontological Security in International Relations
Self-identity and the IR state
Brent J. Steele

The International Politics of Judicial Intervention
Creating a more *just* order
Andrea Birdsall

Pragmatism in International Relations
Edited by Harry Bauer and Elisabetta Brighi

Civilization and Empire
China and Japan's encounter with European international society
Shogo Suzuki

Transforming World Politics
From empire to multiple worlds
Anna M. Agathangelou and L.H.M. Ling

The Politics of Becoming European
A study of Polish and Baltic post-Cold War security imaginaries
Maria Mälksoo

Social Power in International Politics
Peter Van Ham

International Relations and Identity
A dialogical approach
Xavier Guillaume

The Puzzle of Politics
Inquiries into the genesis and transformation of international relations
Friedrich Kratochwil

The Conduct of Inquiry in International Relations
Philosophy of science and its implications for the study of world politics
Patrick Thaddeus Jackson

Arguing Global Governance
Agency, lifeworld and shared reasoning
Edited by Corneliu Bjola and Markus Kornprobst

Constructing Global Enemies
Hegemony and identity in international discourses on terrorism and drug prohibition
Eva Herschinger

Alker and IR
Global studies in an interconnected world
Edited by Renée Marlin-Bennett

Sovereignty between Politics and Law
Tanja Aalberts

International Relations and the First Great Debate
Edited by Brian Schmidt

China in the UN Security Council Decision-making on Iraq
Conflicting understandings, competing preferences
Suzanne Xiao Yang

NATO's Security Discourse after the Cold War
Representing the West
Andreas Behnke

The Scandinavian International Society
From Norden to the northern dimension?
Laust Schouenborg

Bourdieu in International Relations
Rethinking key concepts in IR
Edited by Rebecca Adler-Nissen

Making Sense, Making Worlds
Constructivism in social theory and international relations
Nicholas Greenwood Onuf

World of Our Making
Rules and rule in social theory and international relations
Nicholas Greenwood Onuf

Maritime Piracy and the Construction of Global Governance
Edited by Michael J. Struett, Jon D. Carlson and Mark T. Nance

European Integration and Postcolonial Sovereignty Games
The EU overseas countries and territories
Edited by Rebecca Adler-Nissen and Ulrik Pram Gad

Power, Realism and Constructivism
Stefano Guzzini

Justice, Order and Anarchy
The international political theory of Pierre-Joseph Proudhon
Alex Prichard

War in International Society
Lacy Pejcinovic

The United States and Great Power Responsibility in International Society
Drones, rendition and invasion
Wali Aslam

The Dao of World Politics
Towards a post-Westphalian, worldist international relations
L.H.M. Ling

International Orders in the Early Modern World
Before the rise of the West
Edited by Shogo Suzuki, Yongjin Zhang and Joel Quirk

International Orders in the Early Modern World

Before the rise of the West

Edited by Shogo Suzuki, Yongjin Zhang and Joel Quirk

Routledge
Taylor & Francis Group
LONDON AND NEW YORK

First published 2014
by Routledge
2 Park Square, Milton Park, Abingdon, Oxon OX14 4RN

and by Routledge
711 Third Avenue, New York, NY 10017

Routledge is an imprint of the Taylor & Francis Group, an informa business

British Library Cataloguing in Publication Data
A catalogue record for this book is available from the British Library

Library of Congress Cataloging in Publication Data
A catalog record for this book has been requested

ISBN 978-0-415-62628-6 (hbk)
ISBN 978-1-315-88843-9 (ebk)

Typeset in Times New Roman
by Taylor & Francis Books

MIX
Paper from responsible sources
FSC www.fsc.org
FSC® C013604

Printed and bound by CPI Group (UK) Ltd, Croydon, CR0 4YY

'Nobody forgets a good teacher.'
This book is dedicated to our teachers who inspired our interest in social sciences:
Keith Hodgson (UK)
Ichikawa Shigeo (Japan)
Iijima Eiichi (Japan)
Benedict Kingsbury (New Zealand/US)
Alaistair Leonard (Australia)
Adam Roberts (UK)

Contents

Contributors

Ayla Göl is Senior Lecturer at the Department of International Politics, Aberystwyth University. She was the inaugural John Vincent Visiting Fellow at the Department of International Relations, Research School of Pacific and Asian Studies, Australian National University in 2002, and a visiting scholar at the Centre for Islamic Studies, University of Cambridge in 2009. Dr Göl has published in leading journals such as *Nations and Nationalism*, *Third World Quarterly*, and *Critical Studies on Terrorism*, and is the author of *Turkey Facing East: Islam, Modernity and Foreign Policy*, which is forthcoming with Manchester University Press.

Charles Jones is Emeritus Reader in International Relations and Director of the Centre of Latin American Studies at the University of Cambridge. His leading research interests have been the modern history of Argentina, international business history, the history of thought about war and trade, and military ethics, especially as expressed in literature and film. Leading publications include *More Than Just War* (2013), *American Civilization* (2007), *E.H. Carr and International Relations* (1998), *International Business in the Nineteenth Century* (1987), and *The North-South Dialogue: A Brief History* (1983).

Richard Little is Emeritus Professor at the University of Bristol. He is a former editor of the *Review of International Studies* and former chair of the British International Studies Association. He is the author of *Intervention: External Involvement in Civil Wars* (1975), *Global Problems and World Order* (with R.D. McKinlay, 1986), *The Logic of Anarchy* (with B. Buzan and C.A. Jones, 1993), *International Systems in World History* (with B. Buzan, 2000), and *The Balance of Power in International Relations* (2007). He is a Fellow of the British Academy.

Iver B. Neumann is Montague Burton Professor in International Relations at the London School of Economics, but wrote this chapter when he was Director of Research at the Norwegian Institute of International Affairs (NUPI). He has written two books on Russian–European relations, and is presently working on a duograph about the Eurasian Steppe with Einar Wigen.

Joel Quirk is Senior Lecturer in the Department of Political Studies, University of the Witwatersrand, South Africa. He is the author of *Unfinished Business: A Comparative Survey of Historical and Contemporary Slavery* (2009), and *The Anti-Slavery Project: From the Slave Trade to Human Trafficking* (2011). His work has also recently been published in *Human Rights Quarterly, Review of International Studies* and *Social & Legal Studies*, amongst others. Dr Quirk also serves as a Rapporteur to the International Scientific Committee of the UNESCO Slave Route Project.

David Richardson is Professor of Economic History and former founder and Director (2004–12) of the Wilberforce Institute for the study of Slavery and Emancipation, University of Hull. He was co-author with David Eltis of the prize-winning *Atlas of the Transatlantic Slave Trade* (2010).

Shogo Suzuki is Senior Lecturer at the Department of Politics, University of Manchester. He is the author of *Civilization and Empire: China and Japan's Encounter with European International Society* (2009), as well as several articles on International Relations theory and East Asian foreign policy that have appeared in *European Journal of International Relations, The Pacific Review, Pacific Affairs* and *Third World Quarterly*.

Darshan Vigneswaran is Assistant Professor at the Centre for Urban Studies and Department of Political Science, University of Amsterdam. He is also a Senior Researcher at the African Centre for Migration and Society, University of the Witwatersrand, South Africa. He has authored *Territory, Migration and the Evolution of the International System* (2013), co-edited *Slavery, Migration and Contemporary Bondage in Africa* (2013), and published articles in journals including *Political Geography, Review of International Studies* and *Policing & Society*.

Yongjin Zhang is Professor of International Politics at the School of Sociology, Politics and International Studies, University of Bristol. He has published widely on topics ranging from International Relations theory to Chinese history, politics, economic transformation and international relations. He is the author of several books, including *China in International Society since 1949* (1998), and *China's Emerging Global Businesses* (2003), and his articles have appeared in *Review of International Studies, European Journal of International Relations* and *Journal of Contemporary China*.

Preface

This volume contributes to the burgeoning literature challenging the Euro-centric and ahistorical character of contemporary International Relations (IR) as a discipline. It explores international relations in the early modern era, before the rise of Western dominance in many parts of the world, examining 'patterns of cross-cultural exchange and civilizational encounters' among 'the rest' and between Europe and 'the rest' (even this terminology highlights the extent of the problem!). Through a series of case studies in East Asia, the Middle East, Africa, Latin America and Russia, the empirical chapters demonstrate the importance of non-European agency in the development of global history as they decentre the Europe of the Wesphalian myth.

More specifically, the volume shows that, and how, in many cases the non-European world set the terms of engagement between different regions, including Europe, such that various regional international orders were constructed in significantly different ways. It also highlights how European elites often found it difficult to impose their preferred forms of engagement on their counterparts elsewhere. The book thus documents and analyses a variety of regional international orders primarily defined by local interests, agendas and institutions, with European 'interlopers' often playing a lesser role.

This corrective to contemporary IR is urgently needed, not the least because in this Eurocentric and ahistorical form, IR is poorly equipped to understand and to theorize both the (purported) decline of the West and the rise of non-European parts of the world, such as China. As the editors argue, theories such as the English School and constructivism, while challenging the neo-neo theories in important ways, have continued to reproduce the Western meta-narratives about the fundamentally European nature of international relations and the fundamentally European nature of central IR concepts, including sovereignty, modernity and development.

The central problematique of the volume is thus the dualism of 'the West and the rest' in the current discourse on the emerging global order. As the editors argue in their Introduction, the 'belief that the rise of the non-West is a *novel* event, thus an unknown quality in the history of the evolution of the international system is … highly problematic for two main reasons.' First, as they note, it is historically inaccurate: for most of human history, peoples geographically in Europe have shared the global stage with powerful actors from other continents. Second, the Westphalian myth constructs Europe as the 'norm', as if global history only begins when the West imposes its dominance over 'the rest' and can dictate the forms of engagement between them. This, as this volume convincingly shows, 'is clearly false'.

Jutta Weldes
Series Co-editor

Acknowledgements

Like many collaborative projects, the idea for this book first emerged during an international conference. It has taken some time (as is the case for many projects that owe their origins to bar conversations!) to get the project moving and steer it to completion, but this would not have been possible without the help of many individuals and institutions, and it is our pleasure to be able to thank them here. First and foremost, we thank all our contributors, who gave much of their time to answer our editorial queries and respond to our requests for revisions: your patience has been much appreciated. Earlier drafts of these chapters were presented at the Annual Conventions of the International Studies Association and at a writers' workshop at the University of Bristol. We are grateful for all participants who gave valuable feedback and suggestions. We would like to acknowledge the generous financial support provided by the Vice-Chancellor's Development Fund of the University of Bristol with the workshop. We are also indebted to Routledge, particularly Alexander Quayle, for his patience and good humour. As always, we thank our friends and family, particularly Hirono Miwa, Jiang Yang, Kitamura Akihiko, Arthur Mühlen-Schulte, Len Seabrooke, Stacey Sommerdyk, Jan Sturm, Catherine and Toshiaki Suzuki, Eleni Tsingou, Wan Shanping, and Duncan and Angela Wigan, who provided us with friendship, laughter and support. Finally, we would like to express our gratitude to our teachers from all over the world. Now that we have become teachers ourselves, their dedication to education and ability to inspire their students has made them even more important role models to us. It is in this spirit that we dedicate this book to our teachers who first sparked off our interest in social science and helped us – even though they may not have been aware of it at the time – become who we are today.

Shogo Suzuki, Yongjin Zhang and Joel Quirk
Copenhagen, Bristol and Johannesburg
On Chinese New Year, 2013

Introduction

The rest and the rise of the West

Shogo Suzuki, Yongjin Zhang and Joel Quirk

This book sets out to challenge the Eurocentric foundations of modern International Relations (IR) scholarship by examining international relations in the early modern era, when European primacy had yet to develop in many parts of the globe. Through a series of regional case studies on East Asia, the Middle East, Africa, Latin America, and Russia, this book explores patterns of cross-cultural exchange and civilizational encounters, placing particular emphasis upon historical contexts when 'the rest' often set the terms of engagement and encounter with the West and where European elites found it difficult to impose their preferred modes of interaction upon their counterparts in other continents. The chapters of this book document and analyse a series of regional international orders that were primarily defined by local interests, agendas and institutions, with European interlopers often playing a secondary role.

These perspectives emphasize the central role of non-European agency in shaping global history, and stand in stark contrast to conventional narratives revolving around the 'Rise of the West', which tend to be based upon a stylized contrast between a dynamic West and a passive and static East (the rest). By focusing on a crucial period of global history that has been neglected in the field of International Relations, the book reveals profound differences between the early modern era and the more familiar colonial conquests of the second half of the nineteenth century. Its fresh focus on cross-cultural exchanges and cross-civilizational interactions in contexts of relative European weakness has important implications for understanding and exploring possibilities for interactions between diverse civilizations today. The central problematique of this book is the dualism of the West vis-à-vis 'the rest' in the current discourse on the emerging global order.

The West vis-à-vis the rest as a problematique

For too long, International Relations scholars have taken as unproblematic 1500 and 1648 as two 'orthodox benchmark dates' in the studies of the history of world politics (Buzan and Lawson 2012b). Columbus's discovery of the Americas in 1492 and Vasco da Gama's landing on Culicut in 1498 have been conventionally taken as the beginning of the European expansion to conquer the non-European world. The peace of Westphalia, on the other hand, has been frequently associated with the origin of the 'international', the modern system of states, and the sovereign order in world politics. Taken together, it is often claimed, these two watershed dates mark the beginning of the era of European dominance of the world, political, military and economic, which gave the material, institutional, normative shape and the constitutional foundation to contemporary international society (Bull and Watson 1984a; Watson 1992; Philpott 2001). In both teaching and research in IR, the construction of an essentialized historical narrative of half a millennium of Western dominance underlined by European power is for good reason generally accepted and even celebrated, rather than critically reflected upon and interrogated. As Buzan and Little (2000: 241) note, 'Implicitly nearly all [IR] theorizing is based on the world after AD 1500 and especially after 1648'. Pervasive in the existing IR scholarship is unsurprisingly the dualistic conception, either explicit or implicit, of the West vis-à-vis 'the rest' in the global power and normative hierarchy as a dominant dynamic in international relations. Emblematic of such Eurocentric conceptions of global politics in its contemporary manifestation is Samuel Huntington's (1993, 1996) claim of the clash of civilizations.

The so-called Westphalian myth and the associated narratives have been under sustained and relentless assault in the recent production of IR scholarship (Spruyt 1994; Osiander 2001; Teschke 2003; Hobson 2010; Kayaoglu 2010; de Carvalho *et al.* 2011). It is the discourse of global power shift, however, that has done much to begin to unravel, perhaps unwittingly, the essentialized grand historical narrative of Western dominance since 1500. The hubris of power exemplified by the unipolar moment is followed by its intellectual nemesis: the claim of the terminal decline of the American power, and more broadly of the West (Mahbubani 2008; Jacques 2009; Kennedy 2010; Rachman 2011; Layne 2012). The anxieties of the present, i.e. the shift of both power and wealth from the West to the non-West in the twenty-first century, which claims to have spelt the end of much acclaimed Pax Americana and Western dominance, have returned history to the imagination of the international. Niall Ferguson's (2011) claim of the Great Reconvergence of the West and the rest after hundreds of years of the Great Divergence, Ian Morris's (2010) penetrating question why the West rules for now, and Charles Kupchan's (2012) contention of a new global turn leading to no one's world all try to historicize the rise of the West on a global scale and in a very deep historical perspective from 500 years to 1,500 years. The lens of the *longue durée* in world history has been productively employed to shed new light on the rise the West in order to understand its allegedly irrevocable fall. Such historicization may not necessarily resonate with other critiques of Eurocentrism in world history and IR. However, it helps to reinforce some earlier but marginalized efforts at problematizing the Eurocentric reading of history in articulating the regional and the global in the making of modern world politics and global economy. In this instance, one could look meaningfully at how China, through export trade, functioned as the 'ultimate sink' of the world's silver (Frank 1998: 127), which 'sucked the Europeans into directly joining the global economy' (Hobson 2009: 681) in conjunction with Takeshi Hamashita's (2008) study of Asia's domination of the global circulation of silver in the sixteenth to the nineteenth centuries. One could point to statistics provided by Angus Maddison (2007a, 2007b) to interrogate the claim of European dominance in the global economy before the nineteenth century and to substantiate the contention that 'East Asia remained in the forefront of world development' in the sixteenth to the eighteenth centuries (Arrighi *et al.* 2003: 3).

If history is indeed moving on, a subterranean Eurocentrism/European exceptionalism still dies hard. The seemingly perpetual dualism of the West vis-à-vis the rest is obdurately entrenched in both academic literature and popular discourse from G. John Ikenberry's (2008) speculation on the rise of China and the future of the West, to Fareed Zakaria's (2009) envisioning of a post-American world with the rise of the rest, and from Niall Ferguson's (2011) claim of the Great Reconvergence of the West and the rest, to Charles Kupchan's (2012) conjecture of the coming global turn. The West continues to be associated with progressive agency, rationality, order and sources of (universal) values and norms that constitute and sustain the global liberal order, whereas the rest, the caricatured non-West, continues to be cast as the mirror image of the West with regressive properties and at the receiving end of the benefits of Western civilization. Robert Kagan (2008) reinvented this dualism as the axis of democracy (the West) vis-à-vis the association of autocrats (the non-West). History has returned in the aftermath of unipolarity, with, in his words, 'the old competition between liberalism and autocracy has also re-emerged, with the world's great powers increasingly lining up according to the nature of their regimes' (Kagan 2008: 4). The clash of civilizations, in various incarnations, casts a long shadow.

In this way, the discourse of global power shift harks back to political and intellectual debates over the last two decades on issues of culture, religion and the 'clash of civilizations'. These issues often have unsettling implications for a privileged minority in the West, reflecting widespread unease surrounding the increasing profile of radical Islam, unwelcome migrants, and the rapid and potentially 'destabilizing' growth of Asian powers such as India and China. One of the main points at issue here has been the extent to which different cultural and religious groupings can sustain stable, respectful and peaceful relationships. This contentious topic has both domestic and international dimensions, with the former revolving around concerns about social cohesion and cultural recognition, and the latter revolving around concerns about the operation and organization of a global international order populated by many cultural communities. This frequently translates into a suspicion of the political agency of non-Western peoples and anxieties and concern about the possibility that a fragile yet longstanding and highly valuable international order would be potentially imperilled by the increasing autonomy and influence of political actors who lack a solid grounding in the mores of European civilization. In such light, the rise of the rest cannot be but inherently frightening and potentially destabilizing.

The belief that the rise of the non-West is a *novel* event, thus an unknown quality in the history of the evolution of the international system is, however, highly problematic for two main reasons. First, it is historically inaccurate. The much-touted emergence of a plural international order where non-European peoples have acquired greater political power, influence and agency is simply a reversion to a longstanding pattern. For the vast majority of human history, peoples in Europe shared a global stage with powerful actors of various stripes. Until the mid-nineteenth century, powerful empires in the Middle East, the Indian subcontinent and East Asia were consistently at the forefront of human civilization. While peoples in Europe regularly made important contributions, most major centres of cultural, technological and political excellence were located in other regions of the globe. Before the nineteenth century, 'the European system of states was a chaotic and peripheral component of a global economy that had long been centred on Asia' (Arrighi et al. 2003: 259). That is to say that in deep world historical perspective, the dominance of the non-West was in fact the norm. In this light, the global domination of the West in the last two centuries, particularly since the age of high imperialism, is perhaps more appropriate to be seen as an 'aberration' in world history (Mahbubani 2008).

Second, treating Western dominance as the historical 'norm' betrays a profoundly Eurocentric interpretation of world history, as if international history only began once the Western world was able to impose its dominance over the non-European world and to dictate the terms of interaction with the latter, a process that is frequently assumed to begin after the discovery of the 'New World' in 1492. This is clearly false. As will be examined in chapters in this book, Europeans often had been in contact with non-European political and economic actors much earlier, and their dealings with Asians or Africans after 1492 were often dictated and negotiated on the basis of European weaknesses and limitations. By relegating this particular period of history to the periphery, conventional narratives continue to deny 'Eastern' agency in the making of world history by purposely consigning non-Western peoples to passive, reactive and subordinate positions. The effectively seals off a significant part of the world from the analysis of international history, and contributes to dubious assumptions about how international relations can only be meaningfully conceptualized and conducted within the framework of European political institutions.

Silence and bias in the English School and constructivism

Theories of IR that emphasize the importance of historical context are curiously silent on historical periods and settings defined by non-European dominance. While constructivist and English School approaches have been interested in the historical development of international norms, their focus has been almost exclusively on the global diffusion of social rules and norms that originated from Europe, such as sovereignty, international law or liberal internationalism (e.g. Ruggie 1993; Philpott 2001; Bull and Watson 1984a; Gong 1984a, 1984b; Clark 2005). The context in which this process took place is of course European strength, particularly during the height of European imperialism from the mid-nineteenth to mid-twentieth centuries. Armed with the exceptionalist belief that the white races of Europe were uniquely qualified to spread the tidings of 'civilization', European states were able to impose their own political institutions on other 'barbarous' or 'savage' peoples and polities.

The fact that cross-cultural relations *prior* to European ascendancy hardly feature in these analyses reveals a number of implicit beliefs. First, the fact that 'international relations' governed by shared norms is seen as taking place outside the European world only after *Western* norms are imposed again highlights a deeply Eurocentric belief that the only truly 'international' and 'cosmopolitan' norms that can be shared among humankind are those that originate from the West. Second, the fact that non-Europeans hardly play a role in the production of 'international' norms also assumes that non-Europeans are nothing but 'norm-takers'. They exist only as objects with very little agency waiting to be 'socialized' into norms emanating from the West. Non-Western ideas and norms are thus systematically ignored, and only feature 'to illustrate their incompatibility [with] the Western Westphalian ones that embodies political and religious tolerance' (Kayaoglu 2010: 210), thus reconstructing a narrative of Western moral superiority.

Such a narrow understanding of world history is intimately linked to the normative agendas of both English School and constructivist scholars. The English School approach to IR emerged in the context of the Cold War and the dominance of realism, which posited the existence of a timeless 'anarchical' international realm characterized by insecurity. Keen on discovering possibilities for a more 'moral' international life under anarchy, scholars of the English School embarked on an enquiry that traced back the development of various institutions that emerged in intra-European relations which were held to have eventually evolved into an 'international society'. The fact that this endeavour was intended not only to suggest but also *prove* the possibility of a moral international life meant that international society was seen as something inherently *desirable*, and as such constituted a felicitous achievement of Western civilization (Wight 1966).

As such, the expansion of (European) international society almost axiomatically became *the* key historical period that was worthy of examination. Naturally, the notion that non-European actors were able to construct social rules that could promote some form of coexistence between states did not enter the minds of English School scholars. Indeed, in the English School's narrative of the historical evolution of international society there is little scope for non-European actors to exercise any agency, as they are usually depicted as passive recipients of European norms (called the 'standard of civilization' in the late nineteenth century) (cf. Suzuki 2009). English School scholars' 'firm confidence in the virtues of western civilization' meant that when non-European thought or input featured in their analysis, it was 'accompanied by their fear that [Western civilization] may be undermined by other civilizations (or barbarisms) inimical to its fundamental values' (Suganami 2003: 264), rather than something that could potentially contribute to enhanced order and stability in international politics. Some scholars such as Wight – while never questioning the moral superiority of 'Western values' – did voice their doubts over the possibility that an international society could emerge in the absence of a common culture (Wight 1977: 33), but these concerns were again swept aside in the English School scholars' attempts to 'prove' that a truly 'global' international society had emerged. Thus, one of the most famous works of the English School, *The Expansion of International Society*, is largely concerned with the (decidedly seamless) acceptance of the 'institutions' of European international society by non-European states – namely international law and European diplomatic institutions. This theme was echoed in Gerrit W. Gong's (1984a) *The Standard of 'Civilization' in International Society.*

Similar traits are visible in Ian Clark's (2005) *Legitimacy in International Society* or Edward Keene's (2002) *Beyond the Anarchical Society.* Both authors display strong awareness and sensitivity towards the Eurocentric biases in works by earlier generations of scholars, yet their analyses remain firmly rooted in the history of Europe and its 'rise' to global dominance. Keene makes a valuable contribution to the field by highlighting the coercive, 'civilizing' face of European international society, but focuses exclusively on *European* intellectual debates and history, and does not feature the voices of those at the receiving end of Europe's *mission civilisatrice.* Similarly, Clark's work, which traces the evolution of the norms surrounding the issue of legitimate membership and conduct, is based almost exclusively on European history. Kayaoglu (2010: 207) has critiqued this focus for effectively robbing non-European polities of their agency. '[T]he only time these societies are included in the narrative', he states:

> is to show how these non-Western societies violated the principle of legitimacy and thus are denied membership in international society and sanctioned by Western states … their contribution to the development of international legitimacy is minimal and passive, serving as foil for the core Western states that are the authors of the 'Westphalian' legitimacy.

Constructivists similarly share this shortcoming of allowing their normative agendas to cloud their purported sensitivity to context and historicity. While a sizeable number of constructivists continue to adhere to a stricter positivist notion of social science, it is important to note that there has been a significant strand of contructivist scholarship that has been strongly influenced by Critical Theories of IR (Price and Reus-Smit 1998). In a somewhat similar fashion to the emergence of the English School approach, Critical Theories of IR appeared in the intellectual backdrop of normative discomfort towards dominant IR theories for perpetuating fear and instability in international politics through their uncritical reification of an 'anarchical' international system. Critical scholars charged that existing positivist IR theories' adherence to 'objective analysis' merely analysed and reproduced the status quo, and failed to forward alternative policies and concepts that would lead to more emancipatory outcomes in international political life (Cox 1996). Thus, many constructivist scholars have not only sought to demonstrate that 'norms matter' in international politics, but also prove that *moral progress*, epitomized by the global diffusion of liberal values, is also possible.

The resulting constructivist analysis of 'international history' is decidedly Eurocentric, as it depicts the evolution of the international system as

> the emergence of norms in Europe and then their subsequent diffusion to non-European entities through state socialization. Essentially, international norms like sovereignty, secularism, and human rights emerge from the norm-generating European core, and then diffuse into the norm-receiving non-European periphery.
>
> (Kayaoglu 2010: 209)

Consequently, these perspectives assume 'self-generating Western agency' and 'unwittingly [naturalize] Western civilization and [at worst] Western imperialism' (Hobson 2007a: 93). This viewpoint is visible in the works of Risse, Ropp and Sikkink (1999), who effectively conceptualize the liberal democratic world (dominated by the West) as the 'producers' and 'teachers' of norms that shape other, less progressive actors, into the desired liberal mould. Constructivist scholars' implicit confidence in the progressive qualities of liberal internationalism and its universality means that the spread of this ideology is effectively the only way to conceptualize 'moral progress' in the international realm.

It thus comes as no surprise that both constructivists and English School Scholars ignore the possibility that the history of interactions between Europeans and non-Europeans *prior* to Western domination may have generated norms that promoted some form of shared morality between the two. Indeed, in Daniel Philpott's (2001) account, even decolonization – one of the key developments that punctured European dominance and cultural superiority in the international system – is traced back to European ideas and influences, 'such as the education the colonial elite received in the metropoles or the fact that the metropoles established the institutions and the principles of equality that the Church had embraced', and by using this Eurocentric framework 'non-Western ideas and events in the creation of international society are explained away' (Kayaoglu 2010: 212). There is consequently little room for 'theorizing cross-civilizational and cross-regional interdependencies', and this 'thwarts the accommodation of pluralism' (Kayaoglu 2010: 195). Western liberalism is seen as the 'only game in town', and if emancipation in the international realm is to be achieved, the 'rest' must be homogenized into the norms of the 'West' (Jahn 1998: 637–41).

Such Eurocentric interpretations of history have left IR curiously ill equipped to theorize and understand international relations in a (purported) age of Western decline. While it is argued that non-Western polities' acceptance of the sovereign state system constitutes empirical evidence of the internalization and global diffusion of European-originated norms in international politics, it is important to acknowledge that many non-European states accepted these 'rules of the game' at gunpoint, and could not exercise much choice over this matter. We should avoid overestimating the degree to which many supposedly 'global' norms have attained legitimacy. If the rise of states such as China, India or Brazil will, as it is often claimed, bring about a greater degree of plurality in concepts of international order, then existing theories of IR based on an historical narrative of European ascendance provide us with very few insights as to how we should understand and theorise this new dynamic.

Plan of the book

The working premise of this book is that international orders have never been as cohesive or culturally homogenous as the literature of IR would have us believe, and that there has been a systematic and deeply problematic 'forgetting' and 'remembering' of cross-cultural interactions *before the rise of the West* in the existing literature. We are interested, therefore, in how cross-regional international orders were contested, conceded, compromised, mediated and negotiated in an historical period when the rise of the West was an historically novel event, and what these historical encounters between the rest and the rising West might tell us about the practice of global pluralism in world history. Analytically, our main historical focus is the period *before the rise of the West*. This period can be loosely defined in terms of two historical bookends: 1492 and 1793. The former refers to the discovery of the 'New World', which ultimately paved the way for the emergence of a nascent global order. The latter refers to the first British embassy to China in 1792–3, when Lord Macartney famously refused to kowtow to the Chinese emperor. This incident can be framed as a symbol of a much larger structural transformation, as the Chinese had long been at the vanguard of human civilization, but were now finally being eclipsed by long-term developments within Europe. In this context, 1792 is held to be emblematic of a larger historical trend, as it sits between the gradual decline of the Mughal and Ottoman Empires during the eighteenth century, and the subsequent European conquest of most parts of Asia and Africa during the nineteenth century.

This period can be distinguished from the more familiar 'high imperialism' of the late nineteenth century in other important respects. In the sixteenth century, even after Westphalia, the Atlantic remained the main area of competitive struggles for rival European powers. European intellectuals expended considerable energy discussing the conquest of the Americas. The key point at issue was not whether conquest was feasible, but the extent to which it was moral. These debates have been discussed in considerable depth (Keal 1995,

2003), but in such a way that they are effectively insulated from parallel events in other parts of the world. There were, for example, no equivalent discussions about the morality of conquering the Middle East, since the Ottoman Empire was far stronger than any European power, or plausible combination of powers. A somewhat similar pattern applies in other parts of the globe. Although Europeans established trading relationships in northern and western Africa, conquest was mainly limited to a small number of often tenuous enclaves. In Asia, European envoys were regularly forced to operate within (or around) indigenous models of political authority, as the Mughal and Chinese Empires incorporated traders into existing suzerain systems. The Japanese excluded Europeans for centuries, save a designated port for Dutch traders. In some of these cases, it was indigenous elites who were politically ascendant, and it was Europeans who appeared as supplicants. In other cases, a number of European powers acquired a degree of political and economic significance, yet still fell well short of political dominance. Framed in these terms, the European conquest of the Americas constitutes one end of a political spectrum, rather than a representative example of international order and cross-cultural exchange between the rest and the rising West. 'The millennial balance of the ecumene among the civilizations of the Middle East, India, China, and Europe', in the words of William McNeill (1963: 653), 'did not decisively collapse until about the middle of the nineteenth century'.

Each chapter of this book surveys these issues through case studies of different geographical areas. While they can be read *à la carte*, as independent works in their own right, several common themes emerge. First, throughout the period under examination, trade has performed a powerful role in bringing various polities of often different cultural backgrounds together and aiding the emergence of shared rules and norms (cf. Buzan 2004; Schouenborg 2013: 22–23). This again serves to highlight the empirical (and theoretical) difficulty of sustaining conventional English School scholars' strict differentiation of an 'international system' and 'international society', as well as establishing the precise point at which a 'system' is transformed into a 'society'. Darshan Vigneswaran's chapter, for instance, shows that British expansion into India was an incremental process, often motivated by desires for personal gain by both British and Indian actors, rather than being motivated by the desire to spread the trappings of 'civilization' or judging non-European rulers and polities by the yardstick of 'the standard of civilization'. A similar picture emerges from the Japanese–Dutch case examined by Shogo Suzuki. In their analysis of cross-cultural exchanges in Atlantic Africa, Joel Quirk and David Richardson emphasize that mutual desires for socio-economic gain played a powerful part in facilitating the transatlantic slave trade even before the age of European ascendance. They depart from postcolonial perspectives largely blaming the West for exploitation of African peoples, and highlight how African rulers acted as autonomous actors in their dealings with Europeans (in the somewhat negative sense that they were also culpable for the flourishing of the slave trade across the Atlantic).

Second, the chapters mount a strong challenge to the assumption of a 'Eurocentric endogenous logic of immanence through which Europe's rise is self-generated before it subsequently projects its global will-to-power in order to remake the world in its own image' (Hobson 2007a: 94). For a start, the empirical findings of the chapters show that the Europeans were historically far from the 'dominant actors' that they are often presumed to be, and that there was nothing 'inevitable' about the rise of the West. Suzuki, Zhang, Göl, and Quirk and Richardson all emphasize how Western actors had to work within the socio-political limits imposed on them by the local political and economic actors. In so doing, the authors successfully bring back non-Western agency, which has been subject to much neglect in the field of IR. Furthermore, the chapters historicize and relativize the view of a monolithic, ideologically 'pure' Europe as a 'norm-generating core' from which 'international norms' emanate. Zhang's empirical findings, for instance, challenge the assumption (often visible in previous English School works) that the adoption of norms originating from European international society would help bring about progress and stability in international politics. Zhang suggests that by following Chinese norms of interaction, the Chinese and Europeans seem to have managed a sustained and prolonged period of peaceful co-existence, with only isolated instances of violent conflict between the sixteenth and nineteenth centuries. This stands in sharp contrast to violence, conflict and war between China and the European international society in the mid-nineteenth century, when the latter imposed Westphalian institutions on East Asia by force and coercion.

While not explicitly examining a case of non-European historical ascendancy, Charles Jones provides a critique of the English School's tendency to see the expansion of European international society as a seamless process of European norm diffusion. His case study, which draws on the case of Latin America, paints a complex picture where the native populace, creoles and European colonial rulers coexisted uneasily side by side. The result, he argues, was not the simplistic 'expansion of international society', but the emergence of an *American* international society, where short-term and sometimes ad hoc agreements between the European descendants and the indigenous peoples were the norm. By pointing to these historical complexities, Jones highlights the ahistoricity of the English School (which is contrary to their self-proclaimed sensitivity to history), which results from their belief that (European) normative hegemony is the only pathway to stability and order. Iver Neumann's chapter, meanwhile, demonstrates how the rise of Moscow was accompanied by considerable hybridization of Russian and Mongol practices and culture. The Russians actually took pride in seeing themselves as the descendants of the Mongol rulers. He thus provides historical insights that suggest that so-called 'international norms/culture' are in fact not always influenced by Western civilization alone. Such findings also generate interesting implications for thinking about the possibility for genuinely cross-cultural norms to emerge in contemporary international society. Vigneswaran's chapter suggests that

European actors (in this case the British) displayed very little interest in propagating their own norms. By so doing, Vigneswaran is able to paint a fascinating picture of the 'expansion of Europe' that challenges the English School's notion of an almost seamless diffusion of European customs and norms whenever the 'West' and the 'rest' come into contact with one another.

Revisiting this particular historical epoch when diverse political cultures and contending ideas of domestic and international orders, East and West, competed for recognition and dominance will not only correct a very lopsided, unilinear and myopic view of world history. By giving due consideration to a variety of historical contexts where Europeans were unable to impose their preferred models, all the chapters of this book provide far greater scope and precision to our understanding of non-European agency and non-European institutions, and thereby offer invaluable insights into the contentious relationship between cultural pluralism and international order. We want to demonstrate that the cross-cultural relationships occurring during this period were often far more elaborate than a 'natural law' framework would suggest, and also typically involved extensive input from non-European political elites with little or no knowledge of the vagaries of European intellectual history, thus restoring non-European voices to their rightful place in international history. In this way, critical interrogations of cross-cultural history contribute to theorizing what Kayaoglu (2010: 195) calls 'cross-civilizational and cross-regional interdependencies' in global international society today.

1 Europeans and the steppe

Russian lands under the Mongol rule

Iver B. Neumann

> It was endemic on the medieval religious frontier not to admit consciously that one had borrowed institutions from conquered or conquering peoples of a different religion. This was true of Crusader Valencian 13th century Spain about Islamic Moorish institutions, of the Arab Umayyad dynasty from the 7th century or the Ottoman Empire from the 14th century about Byzantine institutions, and of the French Crusader Kingdom of Jerusalem from the 12th century about Islamic institutions.
>
> (Halperin 2000: 238)

Introduction

The editors write in their introduction to this volume that it was only from around 1750 onwards that European powers acquired sufficient capacities to regularly dictate terms to political communities in many parts of the world. It took that long to muster the capacity to project force across the distances in question. In so far as relations with non-European political communities in Russia were concerned, however, the turning point came 250 years before, in the final decade of the fifteenth century. It was the decade when Russians threw off what they in retrospect chose to name the 'Tatar Yoke'. By 1750, Russia was only decades away from annexing the Crimean Khanate, the last of the other successor polities of what has anachronistically been called the Golden Horde, but that was known at the time as the Khipchak Khanate (Halperin 2000; Morgan 1986: 141).[1]

The annexation followed a victorious Russian war against the Ottoman Empire. At the end of the period covered by this book (1492–1792), then, Russia's relations with non-European polities such as the Ottoman Empire and Persia, not to mention relations with indigenous peoples throughout Siberia, mirrored the hierarchical relations between European and non-European polities discussed in the other chapters. At the beginning of the period, however, Russia was emerging from a clearly subaltern relationship with a non-European polity, namely the Khipchak Khanate, to which Russian cities were suzerain.

Taking its cue from the editors pointing out that the past was not culturally homogeneous, the first part of this chapter is a reminder of the importance of the steppe not only to Russian history, but also to European history at large. The Khipchak Khanate was the polity – the empire, really (Nexon and Wright 2007) – to last the longest of those that came out of the great Mongol empire that ruled most of the known world in the mid-thirteenth century. The Mongol empire was one – it turns out to be the last one – of a succession of polities that, beginning at the end of the third century BC, began its life cycle in the Altai in the extreme north-east of the Eurasian continent, only to absorb a number of other Mongol and Turkic nomadic elements. These were to be found throughout the steppe, which stretched from the Pacific in the south to the forested areas at the Dnepr in the west, and was delineated by the taiga in the north and by sedentary cultures to the south (principally China, Persia and various iterations of the Roman empire) (Barfield 1989).

The second part of the chapter focuses on how Russian cities experienced being part of a Mongol polity, and the third part on how the consequent hybridization fed into Russia's entry into international society. The theme of this chapter, then, is the experience by a specific sub-set of Christians (we cannot in good faith call them Europeans, because this was a concept that was in use briefly during Charlemagne's reign and did not pop up again until the first half of the fifteenth century) of being dominated by 'non-Europeans', and how this experience was, in the period of key interest to this volume (1492–1792), somehow seen as contaminating by other Europeans.

Note that, as seen from the Khipchak Khanate, the key point of interest was not Russia or Europe, but first the Mongol imperial centre at Karakorum, and when the centre lost its hold towards the end of the thirteenth century, another of the Mongol empire's successor states, namely the Il-khan empire to their south-east, with which it quarrelled continuously about tribute taking in and trade routes through the Caucasus (of which more below). Mongols were past masters of many things, one of them being to provincialize the Russian lands. Western historiography has, nonetheless, insisted on treating what happened in these parts, at this time, from a Slav, sedentary point of view only. We may see this already in the name given to the polity, which is 'the Golden Horde'. That was how Russian speakers, a subaltern part of the polity, referred to it. The narratives about 'the Golden Horde' are predicated on 'the notion that the West properly deserves to occupy the centre stage of progressive world history, both past and present' (Hobson 2004: 2). The concept for such a stance is Eurocentrism. The point has often been made that Russian historiography has tried to bury the length and intensity of Mongol control over Russian lands. European historiography has been little different.

The Mongols

The size of the Mongol population at the time of Chinggis Khan has been estimated at 700,000 (Allsen 1987: 5). Although the Mongol made eminent use of heavy wooden saddles and composite bows, their key advantage in warfare was their strategy. The Mongols emphasized protracted training, advance planning, multi-strand coordination and tight discipline. Alone at the time, they concentrated their thinking not on the single combatant or on a small group of soldiers, but on the *tümen* (Russian: *t'ma*), a unit ideally composed of ten thousand men. It was officially recognized that actual *tümen* would be undermanned, for an 'upper *tümen*' was stipulated as having a minimum of 7,000 troops, a middle 5,000, and a lower only 3,000 (Allsen 1987: 193). The land needed to man a *tümen* was also used by the Mongols as the basic administrative unit.

In Europe, Mongols are sometimes (and in Russia, always) referred to as Tatars. We do not quite know why this is so. According to Matthew Paris, a contemporary who wrote interestingly about how Europeans reacted to Mongols, it was the French King Louis XI who punned that the Mongols, who had almost exterminated a neighbouring tribe called the Tartars, emanated from hell (Lat. *Tartarus*), hence Tatars (Morgan 1986: 57).

The key models on which Chinggis Khan organized his Mongol (or Tatar) empire were those of the Uigurs and the Khitans. The Uigurs, a neighbouring people who were first to be enrolled in the burgeoning empire, were a nomadic-turned-sedentary people who had considerable experience in ruling sedentary populations and cities. The Mongols borrowed their alphabet (and used it until about a century ago), their way of setting up a chancery and the concept of scribes. The Khitans were a semi-nomadic Turko-Mongolian people who had conquered the Chinese in the ninth century, established the Liao dynasty, been displaced, and returned as a key steppe force of the twelfth century. The Khitans, who were brought into the Mongol fold in 1218, had administered a loose and non-confessional steppe empire based on tribute extracted by decimally organized cavalry (Morgan 1986: 49).[2] For this, they had used intermediaries, and these are the direct predecessors of the *darugha* used by the Mongols, the Turkish concept for which is *basqaq* (Morgan 1986: 109). The Mongol intermediaries who ran the Khipchak Khanate in Russia in the early decades were locally known as the *basqaqi*.

The Mongols themselves were almost uniformly illiterate, but they kept written records, which were usually penned by personnel taken from conquered peoples. Except for their famous 'Secret History', though, there is very little by way of Mongol historiography. For obvious reasons, the sedentary peoples whom they conquered have tended to treat them as the Other and give them bad press. Throughout the first half of the thirteenth century, Mongols had a very clear and explicit sense of self. They also had a political ideology, complete with scrupulous rules for how to deal with other political entities.[3] The key idea was that of a heavenly mandate. Knowing that all the steppe empires from the Xiongnu (Huns) in the second century BC onwards had adhered to the same principle of legitimacy, and given that the Xiongnu evolved it concurrently with Chinese imperial ideology (Barfield 1989; de Rachewiltz 1971: 104), we already have the outline of the principle's genealogy. Nothing has only one origin, however, and in this case, too, there may have been a fair amount of hybridization when it came to the Mongols' idea that they were on a universalizing mission from God, seeing that the steppe, Chinese, Islamic and Byzantine traditions all entertained similar ideas and all proselytized in the areas where Mongols lived. It is a stock-in-trade of world history that conquering polities experience themselves as being on a mission from God, and that such a stance makes them rather less accommodating interlocutors than polities that do not understand themselves in this way. As Spuler (1965a: 5) puts it:

> Some contribution was no doubt also made by Christian theories of an oecumenical church under a single central leadership, since certain Mongol tribes had for about two centuries been firm adherents of Nestorian Christianity and had thus had access to Christian thought. Insofar as inferences can be drawn when direct evidence of contemporary political ideas is lacking, it would seem that a peculiar metamorphosis of Christian doctrinal theories into political notions had considerable importance in the development of the Mongol concept of world empire.

The *locus classicus* for this discussion is Voegelin (1941: 402), who analysed the preambles to extant orders of submission from Mongol khans to European powers as 'legal instruments ... attached to the orders of submission in order that the addressees might not plead ignorance of Mongol law when they did not obey the orders received'. Voegelin (1941: 378) extracted from this material 'the principal ideas underlying Mongol constitutional law, as well as the framework of Mongol political theory'. The key idea is the isomorphism between heaven and earth; the former is ruled by a deified Eternal Heaven (*Möngke Tenggeri*), and the latter should be ruled by his servant, the Mongol 'emperor' (khagan). There was, however, a temporal problem, for:

> The true essence of world government is not yet in an actual but only in a potential state, and it is bound to materialize itself in the course of history by turning the real world of political facts into a true picture of the ideal and essential state as visualized by the Order of God ... the Mongols, therefore, cannot simply make war on foreign powers, since any legal title is lacking for an enterprise of this sort. The proper mode of procedure for the Imperial Government is to send embassies in due form to the powers in question, giving them all the necessary information on the principles of Mongol World-Empire law in order that they may know that the moment of passing from potential to actual membership has come, and to enable them to take this step in accordance with the legal rules which govern it.
>
> (Voegelin 1941: 403, 405; comp. Allsen 1987: 42)

In other words, the khagan (as well as his blood relatives, known as the Golden Kin) was fully aware that there were rulers who did not yet know of his existence, but these were nonetheless classified as being in rebellion against the Mongol empire.

Chinggis Khan had four sons who all left descendents: Jochi, Chaghadai, Ögödei and Tolui. Relations between these four lineages were at the centre of Mongol politics. The key principle of organization was kinship, both biological kinship and classificatory kinship. The language of the fights over succession was the one of the *jasagh*, the rules of the ancestors, which were supposed to be upheld and to which respect should be paid, not least when these used were used creatively. Although the custom was for the youngest son to follow his father, when it came to being the khan of khans (khagan), there was no automatic succession involved. The candidates built alliances, which felt one another out until one candidate emerged as the stronger one and called a *kurultai* where the leading Chinggisid successors were to consecrate him (Allsen 1987: 34). After Chinggis Khan died in 1227, his youngest son Tolui took over as regent, but in 1229 it was Ögödei who was made khagan. When he died in 1241, a protracted fight between the Toluids and the Ögödeians ended when Tolui's oldest son Möngke was made khagan in 1251.[4] This protracted fight was of key importance to European history, and I will return to it below.

Centralization of the empire peaked under Möngke. Within his central administration, he established regional secretariats for China, Turkestan, Persia and, although this is not altogether clear, Rus' (Allsen 1987: 101). He recalled all the imperial seals, insignia and orders from the court (*jarligh*) and issued new ones. This gave him a chance to screen all the empire's middlemen and all his own residents. He then restricted the availability of the vital postal system to these people only.

> A third measure was intended to circumscribe the power of the imperial princes within the confines of their own appanages (*fen-ti*). Thenceforth, these princes could neither summon their subjects on their own authority nor issue any orders concerning financial matters without first conferring with officials of the imperial court.
>
> (Allsen 1987: 80–81)

Möngke dispatched his own people to do the actual tax collection. The local middleman was allowed to have his own representative on the spot, but he was not allowed to receive the actual taxes. Allsen (1987: 46) notes that:

> Of particular importance was the qaghan's [khagan] right to appoint the Mongol residents, called *darughachi* or *basqaq*, who were stationed in all major population centers and at the courts of all local dynasts. These officials, who commanded wide administrative, police, and military powers, were key figures in the control and exploitation of the subject populations.
>
> (Allsen 1987: 46)

A final point that needs underlining in our regard is that 'The grand qan [i.e. khagan] had exclusive right to conduct relations with others on behalf of the empire' (Allsen 1987: 45). I have dwelt on Mongol administration and its historical precondition, first, in order to demonstrate that the Mongols stood in a long political steppe tradition and, second, because this was the blueprint for the Mongols who settled on the Volga to rule the Rus' lands from the 1240s on.

The Mongols' western campaign

When Chinggis Khan died in 1227, he had not only instructed his sons to conquer the world, but he had allotted parts that were not yet conquered.[5] The extreme west of the Mongol empire was the preserve of Jochi, who was also bequeathed 4,000 soldiers (Tolui inherited the lion's share, 101,000 men). Jochi had already reconnoitred the lands, and established a fledgling polity we know as the White Horde somewhere north of the Caspian Sea. Indeed, in his work on Mongol imperialism, Thomas Allsen maintains that the 1237–40 expedition that established the Mongols in the Rus' lands 'was designed primarily to carve out a territory for the family of Jochi' (Allsen 1987: 28, comp. 45).

Jochi's reconnoitring in 1223 had also resulted in first contact between Mongols and the Rus'. On their way westward, in 1222, the Mongol reconnoitring party met opposition from an alliance of Alan and Khipchak troops.[6] When the Mongols proclaimed themselves the blood brothers of the Khipchaks, this was enough to break the alliance. The Mongols proceeded to massacre the Alans while the Khipchaks stood idly by. Once the job was done, the Mongols massacred the Khipchaks. The Khipchak Khan Kotyan passed words of what had happened back to his son-in-law Prince Mstislav of Galicia (note the marriage alliance), who called a council in Kiev. Three princes decided to raise an army and engage them on foreign territory. The army marched east, where they were met by Mongol envoys whose message was that their real quarrel was with the Khipchaks. The Rus' princes recognized the tactic that they had heard about from the Khipchaks themselves, and proceeded to kill the envoys. This move guaranteed that there would be war. When it broke, the three Rus' princes were neither willing nor able to coordinate their efforts (which also meant that they could not coordinate very well with their Khipchak allies).

The importance of Mongol superior strategy was in evidence already during this first clash between the Rus' and Mongols, which took place at the Kalka river (now in southern Ukraine) in 1223, when two of Chinggis Khan's four key generals, Jebe and Subudai, outmanoeuvred a badly organized assemblage of Rus' and Khipchak forces which actually outnumbered the Mongols (Allsen 1987: 6). Note that the western reconnoitring played out according to standard Mongol operating procedures:

> Prior to the commencement of hostilities with a foreign state (*qari-irgen* [i.e. polity]) the Mongols always issued orders of submission that offered

> its ruler physical and institutional survival in return for acknowledging the suzerainty of the qaghan [i.e. khagan]. Even if the ruler did not in the end surrender, such offers were still a valuable means of weakening an enemy's resolve and a diplomatic tool for detaching his clients and allies … Another and perhaps more compelling reason for the toleration of dependent states was the Mongols' lack of experienced administrative personnel. Inasmuch as very few of the Mongols' estimated population of seven hundred thousand were literate and still fewer were familiar with the 'customs and laws of cities,' retention of a local dynasty and its attendant administrative apparatus was frequently the most practical method of controlling and exploiting the population and resources of a newly surrendered territory.
>
> (Allsen 1987: 64–65)

The Rus' princes, seemingly reckoning that the Mongols were simply another steppe nuisance, paid no more heed to steppe affairs than before. That was a key mistake. In 1238, the Mongols returned with a vengeance. For the next two years, they effectively overcame all military opposition from Bolgars, Khipchaks, the Rus', Poles and Hungarians. They established themselves in the Rus' and Hungarian lands, and had scouting parties as far west as Venice and Vienna. Once again, the campaign went according to plan. Cities that did not offer resistance were spared, cities that did were more or less destroyed. The result, here as elsewhere in the empire, was patchy destruction of the conquered areas (Morgan 1986: 82).

There is no reason whatsoever to assume that if they had forged ahead, the Mongols would not have subdued all of what we may anachronistically refer to as Europe and made it into part of the Mongol order in one way or another. As it happened, however, news of Ögödei's death reached the extreme west of the empire in 1241. At this time, not only Batu, who was Jochi's oldest son, but also Ögödei's oldest son Gülüg and Tolui's oldest son Möngke were there. The presence of three out of four Chinggisid lineages was not by chance; the western front was at this time the key area of new conquest, which meant that representatives of the different lineages were there to keep an eye on one another. Now, however, it became more important to keep an eye on one another in the Mongol heartland around Karakorum, where the succession would be decided. In the upshot, both Gülüg and Möngke left the western frontier for the steppes. The focus of imperial politics turned away from the fairly narrow strip of land that remained to be conquered, namely Europe. This left the Jochids, led by Batu, alone in the west with his newly won Rus' possessions.

Although he was no longer in the thick of imperial politics, as head of one of the four Chinggisid lineages, Batu remained a key player in Mongol politics.[7] When Khagan Ögödei's widow Töregene, who was regent 1241–46, called a *kurultai* to consecrate Gülüg as new khagan, Batu refused to attend, and when she went on anyway, Batu refused to acknowledge the new khagan.

This was instrumental in forcing the khaganate from Ögödeian hands and usher in the Toluids, and this happened at a *kurultai*, which was actually called by Batu. Furthermore, Batu had more leeway vis-à-vis the imperial centre than had other regional middlemen (Allsen 1987: 61; comp Nexon and Wright 2007). Actually, from Möngke's accession in 1251, 'Batu was conceded virtual autonomy in his own *ulus* [patronage] of the Golden Horde' (Morgan 1986: 117).

Note, however, that the first *darughachi* or governor to the Khipchak Khanate or Golden Horde, a Mongol by the name of Kitai, was sent from Karakorum in 1257 (Allsen 1987: 104). Furthermore, Batu and his immediate successors sometimes sent Rus' princes to the Mongol capital of Karakorum to have their patents of rule confirmed there.[8] Also, under Möngke:

> Hostages were an additional measure designed to assure the fidelity of the Mongols' dependent rulers. Carpini reports that all tributaries were required to send sons or brothers to the imperial court [at Karakorum]. As examples, he notes that Yaroslav of Vladimir, the chieftain of the Alans, and the Korean king had sent relatives as a pledge of their good behavior ... it was not always the possibility of the hostage's execution that kept a dependent ruler in line, but rather the threat of being deposed and replaced by the hostage at the first sign of disloyalty.
>
> (Allsen 1987: 73–74)

When Batu died in 1256, he had built a tent capital in Saray on the Volga (100 km north of today's Astrakhan) for his khanate, which came to be known locally as the Golden Horde. Batu was followed by his short-lived son (Sartaq, a Christian) and grandson, before his brother Berke (1257–66) took over. Berke lost Georgia to another Chinggisid line, the Il-khan of Persia and, as we shall see, this sowed the seeds of discord for centuries to come. However, the overall story of his reign was that he gained more room for manoeuvre within the Mongol empire, the cohesion of which was now definitely loosening (Allsen 1987: 62–63).

Mongols and Rus' polities

The Mongols destroyed Kiev and established a new layer of Mongol overlordship to what was now becoming a suzerain system of Rus' cities within an imperial structure – that of the Golden Horde. The Khipchak Khanate, which was itself still part of an imperial structure, continued to follow the standard operational procedures of Mongol rule. As summed up by Allsen, the basic demands that the Mongols imposed on all of their sedentary subjects were: '(1) the ruler must come personally to court, (2) sons and younger brothers are to be offered as hostages, (3) the population must be registered, (4) militia units are to be raised, (5) taxes are to be sent in, and (6) a *darughachi* is to take charge of all affairs' (Allsen 1987: 114). To the Mongols:

> the surrender of a foreign state [i.e. polity] was not just an admission of military defeat and of political subordination, but a pledge that the surrendering state would actively support the Mongols in their plans for further conquest. To fulfill this pledge, the surrendered state had to place its entire resources at the disposal of the empire, and because a census was needed to identify and utilize these resources effectively, the Mongols came to consider submission and the acceptance of the census as synonymous acts.
>
> (Allsen 1987: 124)

A census was made of Kiev in 1245 and of Novgorod in 1259.

Following Mongol standard procedures, the khan initially dispatched *basqaqi*, personal representatives, to live in key Rus' cities. After some decades (how many exactly is not known), the Mongols changed their policy and dispatched representatives who were based in the capital Saray on shorter inspections (*darugi*). The Rus' called these *posoli* (*posol* is still the term for ambassador in Russian). When the *posoli* were not on missions, they worked in the administration in Saray (Halperin 1987: 33). In the degree that there remained a *primus inter pares* amongst the Rus' princes, it was the grand prince of Vladimir. His rule, like that of all princes, was dependent on a Mongol patent (*yarlik*). The principle of personal presence was replayed on the regional level, which meant that Rus' princes journeyed to Saray in person to deliver their pledges of loyalty. The Rus' probably paid their taxes partly in coin, partly in furs. As will be seen, this is a standard way of establishing order that closely fits ideal types of what imperial relations look like (see Nexon and Wright 2007). It is also a reminder that European forms of imperial control as we know them from the more recent past are not unique to Europe, and so cannot be called particularly European. If nothing else, this should be a reminder that the European moment in world history (*c.* 1800–1950) was special in that it was global, but not so special when it came to form. By the same token, one might expect coming imperial orders emanating from rising powers to be fairly similar to the sundry orders we have already known historically.

In Rus' lands, as elsewhere under the Mongols, there was one group that did not pay taxes. That was religious leaders, which in Christian areas meant the clergy. A precondition of this special treatment was Mongol eclectic religious tastes and general tolerance. Exemption from taxes was also a useful political tool, which facilitated breaking in local religious elites to imperial rule. In Russia, as elsewhere, this came in handy.[9] The clergy was, it will be remembered, a force in the squabbling between lineages in Russia, and this squabbling went on unabated after the Mongol invasion. As Fennell (1983: 97) puts it, 'the princes were able to squabble amongst themselves, to manage their own business, to defend themselves against enemies in the west, and even occasionally to interfere in the affairs of their old neighbours in the south'.

The 'Vsevolodskiys', whose struggles converged on the city of Vladimir and its hinterland (Suzdalia), were the main lineage in Russia after the Mongol invasion. After Kiev's fall, it was Vladimir that was the key Rus' city. The

Vsevolodskiys were named after Yuriy Dolgorukiy's son Vsevolod III, whose son Yaroslav's sons included Aleksander Nevskiy and Andrey. They were, not surprisingly, split on the key question of whether to cooperate with the Mongol invader or to cooperate with their neighbours to the west. This was a struggle for keeps, in the sense that the winner would maintain the throne for his direct descendants (primogeniture having become the key principle of succession in the years immediately preceding the Mongol invasion). Furthermore, since Galicia was already attempting to head westwards and the southern cities were increasingly passive politically, it was also a struggle about the entire orientation of what remained the key areas of the Rus' lands and was increasingly becoming the only centre of political gravity between the Khipchak Khanate in the East and Hungary, Poland, Lithuania, the Germans and the Scandinavians in the West. It was a centre that was very aware of its dependence on their new Mongol overlords. Between 1242 and 1252, Suzdalian princes made 19 visits to Saray. Four of these visits ended with the princes being sent on to the Mongol capital Karakorum (Fennell 1983: 99).[10] Once again, the imperial order on display here, with suzerain supplicating polities quarrelling to get in line, is an example of a pattern that we know from both non-European and European orders.

Given Mongolian superior military force, the temptation to embrace the inevitable and to collaborate must have been very strong indeed. The key bandwagoner was Aleksander Nevskiy. Already in the early years of the Mongol invasion, Aleksander had spent the time successfully fighting Swedish detachments (1240, earning his moniker) and German knights (1242). These fights were part of a protracted struggle for mastery over the lands lying between them. When Yaroslav died in 1248, Aleksander was next in line of succession, but it was his younger brother Andrey who seized the throne. Andrey was one of the few Rus' princes to advocate resistance to the Mongols. Nonetheless, in order to hang onto the throne, he needed the patent from the khan, so both he, and eventually his brother Aleksander, made their way first to Saray, and then onwards to Karakorum, where Andrey was confirmed in Vladimir and Aleksander in Kiev. Since Vladimir had been the main prize since the Tatar invasion, Aleksander did not rest content with this decision, and in 1252 he went to the Horde and obtained their help to oust Andrey. Andrey fled to Sweden. Aleksander had managed to put paid not only to his brother Andrey, but also to organized opposition to the Mongols as such. As Fennell (1983: 108) puts it:

> … this was the end of any form of organized opposition to the Tatars by the rulers of Russia for a long time to come. It was the beginning of Russia's real subservience to the Golden Horde … the so-called 'Tatar Yoke' began not so much with Baty's [i.e. Batu's] invasion of Russia as with Aleksander's betrayal of his brothers.

From this time on, the enrolment of Mongol backing became a routine part of internecine struggles. There was nothing new about this: first the nomadic

Pechenegs and then the Khipchaks had been drawn on in similar fashion by the Rus' princes before. Now, once more, the appeal to steppe forces became a key factor in the intensification of direct Mongol control with Rus' political life. There is a causal link between this development and the period of intensified Mongol raids and invasions towards the end of the thirteenth century. At this point, not only were Mongols from the Khipchak Khanate brought in, but Rus' princes who were up against other Rus' princes with Horde backing actually ventured further afield to bring in the backing of Mongol insurgents from the Nogay further south.[11] Rus' princes stood against Rus' princes, each backed by a Mongol ally. We see here a typical example of how an imperial power overlays a balance of power between suzerains.

In 1304, the grand prince of Vladimir died. Three developments brought about a change in politics. First, the princes of Moscow and Tver' emerged as the key players in Rus' politics, among other things as a result of their population increase in the wake of the Mongol invasion, which was again to do with a nice strategic location (with Moscow, in particular, being something of a hub of the river system).[12] Second, among other things because of the now firmly established principle of primogeniture, these princes headed more clearly organized families, which served as a firm power base. Third, the firm wedding between families and cities meant that the territoriality of this power base was now assured in a much higher degree than before.[13] Following decades of struggle between Moscow and Tver', Moscow emerged victorious and Ivan I was granted the title of grand prince of Vladimir by the Mongols in 1328. From Ivan I onwards, Moscow was the emergent centre of gravity of Rus' politics, and the home both of the great prince (who underlined his success by adding 'and of all Russia' to his princely title) and of the Metropolitan. Moscow remained completely dependent on the Mongols, however, to the point that brothers appealed to Saray and even travelled there in order to settle their succession struggles (Halperin 1987: 58). Moscow took its time fighting down Tver' competition. In 1353, Novgorod supported the Tver' bid for the grand principality of Vladimir over the Moscow one by sending envoys to Saray to plead for Tver's case (Halperin 1987: 51). The suzerain balance of power system was still in operation, and that very operation also served to confirm and strengthen the importance of the Mongol overlay.

The grand princes of Moscow kept up their brilliance in playing this game. Whereas Tver' looked West, to the rising power of Lithuania, Moscow stuck to the Mongols of the Khipchak Khanate. This served them well, for they were able to stave off three attacks by Lithuania and Tver' between 1368 and 1372. As summed up by Halperin (1987: 54; for details, see Vernadsky 1953: 207):

> the special relationship between the Golden Horde and Moscow was strengthened in the middle of the fourteenth century, when the Mongols faced a new challenge to their hegemony. Grand prince Olgerd of Lithuania struck deep into the Tatar orbit by bringing both Tver' and Riazan' into his sphere of influence and applying pressure to Novgorod.[14]

> Olgerd's opposition to Moscow was not rooted in principle, and he played politics by the same rules as everyone else. Thus, with the eye on Moscow, he sent a delegation to the Golden Horde to negotiate a rapprochement. The Mongols, however, had decided, logically, to use Moscow as a counterweight to the growing power of Lithuania. The Muscovites were therefore successful in their attempts to undermine the Lithuanian embassy, and the Mongols, in a fine display of political delicacy, arrested the Lithuanian envoys and handed them over to Moscow. Olgerd was compelled to ransom his emissaries from his enemies.

The decisive Moscow victory over Tver' occurred in 1375.[15] In 1478, Ivan III subdued Novgorod. Moscow owed its victory to the superior way in which they had played the alliance game vis-à-vis the Mongols compared to other Rus' polities. From this time on, in order to underline how Moscow was changing the suzerain system of Rus' lineages into a polity centred on Moscow, it is customary to refer to this polity as Muscovy. Muscovy was still subservient to the Khipchak Khanate, and would remain so for another 100 years.

In terms of systems logic, the arrival of Lithuania was a major event, since it challenged the suzerain system by adding another possible centre of gravity for Rus' princes. It is true that Lithuania was at first sucked into the suzerain system centring on the Khipchak Khanate's ambit, to the degree that the Khipchak Khanate certainly saw Lithuania as a vassal, and Lithuania itself at some point probably did (comp. Vernadsky 1953: 264). It is also true that the Khipchak Khanate backed Moscow in its war with Lithuania in 1406, and also on subsequent occasions. As the Khipchak Khanate weakened and Moscow as well as Lithuania emerged ever stronger, however, diplomatic relations between the Khipchak Khanate and Lithuania became closer and also less lopsided. Despite certain temporary setbacks such as the Moscow-Lithuanian treaty of friendship of 1449 (a short-lived affair anyway), the Khipchak Khanate and Lithuania were more often than not as one on opposing the rise of Moscow.[16] It was an alliance that did not fulfil the goal for which it was formed, however, for Moscow (which could in turn draw on its good relations with the emergent Crimean Khanate)[17] emerged triumphant, whereas the Khipchak Khanate fell apart. Note, and this is crucial in our context, that the patterns of alliance do not follow religious or cultural lines. The same may be said about the alliance that Muscovy and what was left of the Khipchak Khanate formed in 1502, against the Great Horde, i.e. the polity of nomadic Mongol-led forces on the steppe. Cultural similarity eases contact between polities and may result in alliances, but as is demonstrated here, it may also take a back seat to other concerns.

To sum up, the key political fact in the Rus' lands from 1240 to the end of the fifteenth century was the suzerainty of the Mongols, based in Saray. Rus' princes fought one another, and used Mongol backing as the key power resource in their internecine struggles. The Mongols lent their support to various princes with a view to upholding tribute. They also followed the same policy towards the Rus' princes that they themselves and other steppe peoples had experienced from the Chinese side: they played the Rus' princes off against one another so that no one of them should emerge as a uniting force that could challenge Mongol rule. As the Khipchak Khanate started to fall apart from the mid-fifteenth century onwards, however, Moscow was nonetheless able to emerge as the key political centre, which proceeded to relativize Mongol suzerainty and, using techniques borrowed from the Mongols, unite first the Rus' lands and then the old lands of the Khipchak Khanate (Kappeler 2001). Muscovy seems to have stopped paying tribute to the Khipchak Khanate sometime around 1470, and made an alliance with the eastern part of what was left of it in 1502. Muscovy effectively swallowed its partner, and in 1507, Sigismund of Poland-Lithuania was 'granted' the western part from its last khan. The Khipchak Khanate was no more.

Although the fact that the Khipchak Khanate turned to Islam in the first half of the fourteenth century could not fail to delineate them clearly from a population who was consistently referred to by its writing layer as 'the Christians', a number of hybridizing practices were in evidence. There was intermarriage, but it was to a high degree an elite phenomenon. Spuler (1965a: 86) sums it up as follows in the fashion of high modernity: 'A fair measure of Finnish blood was absorbed into the veins of the Tatar nation, and Russian and Polish captives of both sexes added a certain Slavic element, though the Russian contribution was in all probability still very small in the mid 15th century'. More importantly, after more than 250 years of Mongol influence, there was widespread hybridization on the institutional and practical levels. We have here a key example of the kind of situation that this book was supposed to excavate: non-European domination of Europeans invariably led to cultural hybridization. Equally invariably, it seems, and as pointed out in the epigraph to this chapter, such hybridization is quickly and wilfully downplayed once the power configuration that gave rise to the hybridization in the first place, changes.

How widespread hybridization between Mongols, Russians and others had become in this period is a matter of debate, most recently around the publication of a book by Donald Ostrowski (1998). Halperin (2000: 238) outlines the broad consensus as follows:

> Despite the objections of hypersensitive Russian historians, there is a compelling case that Muscovy did indeed borrow a variety of Mongol political and administrative institutions, including the *tamga*, the seal for the customs tax as well as the tax itself; the *kazna*, the treasury; the *iam*, the postal system; *tarkhan*, grants of fiscal or juridical immunity; and *den'ga* for money. Muscovite bureaucratic practices, including the use of *stolbtsy*, scrolls to preserve documents, and perhaps some feature of Muscovite bureaucratic jargon, may also derive from the Qipchaq Khanate, as well as selective legal practices such as *pravezh*, beating on the shins. Certainly Muscovite diplomatic norms for dealing with steppe

> states and peoples were modeled on Tatar ways. Finally, the Muscovites had no choice but to study Tatar military tactics and strategies, if only to survive by countering them in battle, but the Muscovites also copied Mongol weapons, armaments, horse equipage, and formations.

The consensus does not extend to the administrative system as such. Ostrowski (1998, 2000) sees the major aristocratic organ, the Boyar Duma, as being formatted on the major aristocratic organ of the Khipchak Khanate, and also sees a number of detailed similarities between the lower ranks, but has not been able to garnish much support for this view (Halperin 2000; Goldfrank 2000). Be that as it may, the pride that Russians took in being the key successor of the Khipchak Khanate was evident in the sixteenth-century aristocratic fashion for tracing one's ancestry back to Mongols (Halperin 1987: 113). As we shall see, this Russian identification with its former sovereigns proved to be detrimental to its relations with its European neighbours.

Significance for Russian–European relations

Throughout the Mongol period in Russian history, relations with Western Christendom continued. The Khipchak Khanate cherished trade, and gave privileges to a number of traders. Most of the trade went through the Black Sea, and was handled by non-Mongol servants of the khan. The Mongols were generally very good at acknowledging their limited knowledge of city ways, seafaring and other pursuits foreign to the steppe. Seeing the advantages of trade with the known world, they therefore employed foreign subjects as customs officers, and allowed colonies of Genoese and also Venetian traders along the northern coast of the Black Sea (seeing to it that some ports remained in Tatar hands). From 1365 to 1475, when the Crimean Turks put an end to their presence, the Genoese dominated heavily (Meyerdorff 1981). In the early years, the Khipchak Khanate demanded tribute of the Venetians, but, presumably finding this to be counter-productive, soon rescinded the practice (Spuler 1965a: 399). Trading included a whole gamut of goods, and also slaves.

In the immediate aftermath of the invasion (by the Mongols, of the Russian lands), and despite Aleksander Nevskiy's scepticism of Western powers and Catholicism, Pope Innocent IV nonetheless forwarded a bull to him in 1248 (Fennell 1983: 122, n.15). Rome followed what was going on in the Rus' lands. Note also that Aleksander's ally Metropolitan Kirill established a bishopric in Saray in 1261. The church's presence in Saray secured, among other things, a channel from the Rus' clergy and princes to the Byzantine Empire, which had diplomatic relations with the Khipchak Khanate (the Byzantine emperor married off his daughter to Khan Uzbeg of the Golden Kin in around 1330; Vernadsky 1953: 196). The Khipchak Khanate also received diplomatic envoys from Rome. Even in the immediate aftermath of the invasion, 'trade with the West, either from or via Novgorod and Smolensk, both of which suffered no damage from the Tatars, seems to have been relatively unaffected'

(Fennell 1983: 89). Furthermore, the Khipchak Khanate granted tax exemptions to the Hanseatic League, which continued its brisk trade with Rus' lands via Novgorod (Halperin 1987: 81).

In 1270, the Khipchak Khanate also made a trade agreement with Riga Germans and other Germans, and saw to it that Russian princes did not interfere with it. The main route for this trade followed the 'Tatar road' from Kiev to Lemberg. Kiev, the main Russian town, was 'teeming with Tatar, German, Armenian and Moscow merchants' (Spuler 1965a: 403). In the context of this book, the main point here is that the boot was firmly on the Tatar foot when it came to settle the conditions for and terms of trade here, much as they were where the other European-non-European relations discussed in this book are concerned (see, particularly, the chapters by Göl, Jones, Suzuki and Zhang). Note here that Genoese and also particularly Venetian merchants were able to draw on these experiences when it came to evolving trade with the Ottoman Empire. We have here a reminder of how, in the years before the sixteenth century, the polities of Christendom were simply not particularly powerful agents relations with polities from elsewhere.

If the existence of Russian-speakers and their human status were known to most Europeans, the same could not be said about the steppe-dwelling peoples to their east. Ever since Pope Alexander III's personal physician Master Philip had set sail eastward from Venice in 1177 on his mission to find the alleged Christian kingdom of Prester John, attempts to establish contact had rested on 'a strange combination of Christian and pagan elements … [built on] the legends and myths inherited from the classical world' (de Rachewiltz 1971: 29). When the pope had word of the Mongol invasion some 60-odd years later, his reaction was to send friars with letters asking the Khans to mend his ways and convert to Christendom. The Mongol answers mirrored these messages by insisting that the pope should come and pay his respect to the great khan [i.e. khagan]. Universal claim stood against universal claim (Dawson 1955; Bowden 2009). The envoys to the great khans brought back new information that made for much more detailed representations in the West of people and life in the East.[18] However, when both the Ilkhanate state (Mongol-ruled Persia) and the Khipchak Khanate first converted to Islam and then, later in the fourteenth century, went through periods of internal strife, it affected the possibility for European missionaries and merchants to take the land route through these areas in order to reach destinations further east. As a result, direct contacts between the European continent and the East suffered, and European continental representations of the East were once again dominated by 'dreaming and speculation', as de Rachewiltz (1971: 207) puts it. What this meant was that when Muscovy emerged, Western rulers did not know what to expect. Already in 1481, Emperor Frederick III addressed an appeal on behalf of the Germans in Livonia to Poland and Lithuania, Sweden and the Hanseatic cities about this noted but unknown entity (Halecki 1952: 8). Direct contacts between Muscovy and the Holy Roman Emperor ensued in 1486, after two and a half centuries of Mongol rule.

In the early 1500s, Russians themselves were far from certain about what to make of their Mongol connection. There was a duality in the Russian knowledge production about these relations which goes to the heart of how Russo-Mongol relations are relevant to Russia's entry into the European state system. On the one hand, as has been demonstrated convincingly by Charles Halperin, Russian contemporary sources, both the chronicles paid for by princes and literary genres such as the *byliny* (folk songs), finessed a technique of not touching on the fact of Mongol suzerainty directly. As Halperin (1987: 8, comp. 63) puts it:

> The Russian 'bookmen' (writers, redactors, scribes, copyists) of the Kievan past were accustomed to explaining Russian victories and defeats in skirmishes with nomads as signs of God's pleasure or displeasure with his people. They had never been called upon, however, to rationalize absolute conquest. Instead of confronting the ideologically awkward fact of utter defeat, the bookmen finessed the fact of Mongol conquest by presenting Russo-Tatar relations as merely a continuation of Kievan relations with the steppe with no change of suzerainty involved. Thus the Russian bookmen raised the ideology of silence to a higher level and threw a veil over the intellectual implications of Mongol hegemony.

However, once the Mongols seemed to be a spent force, there was a need to tell a story about Russia's history as having some kind of continuity. A solution that lay close to hand was to forge a new role for the Russian leader as being not only a great prince, but also a tsar. The problem was that the term 'tsar' was a translation into Russian not only of the Greek term *basileus* (i.e. Byzantine emperor), but also of *khan*. The implication of these eponymous translations was that these two entities were treated on a par. Note that the fall of Constantinople is at this point half a century back. There was no longer a *basileus* in Constantinople. The hegemon to live down was the khan in Saray. Vassilian, bishop of Rostov and a close adviser of Ivan III, came up with an answer to this problem, namely to raise the status of Ivan III to that of tsar and so live down the very idea that there was ever such a thing as a tsar in Saray. The link should be that of *basileus* to tsar, and the khan should be treated as nothing but an impostor (see Cherniavsky 1970).

However, there is an interesting split in representations of Muscovite rule here, for as I have tried to demonstrate above, once the domestic work of establishing the basic continuation of Russia's legitimacy as a Christian power was done, Muscovy actually started propping up its claims of being an imperial power on a par with the Holy Roman Empire by invoking its conquests of the successor states of the Khipchak Khanate, notably Kazan' and Astrakhan.[19] In a situation where Europeans knew little of Mongol or even Asian ways (little, not nothing: there had, after all, been continuous contact), Russia chose to base its claims for recognition partly on its Mongol connection. This move flowed from hybridization, and the self-evident way in which

Muscovites performed the move goes to show how this hybridization had become doxic (not least, one would suspect, because there was now little to fear from the former overlords). Muscovy opting for a Mongol *translation imperii* was clearly detrimental to the Russian polity's relations with Europe, where Mongols were remembered as clear-cut barbarians. Even as late as in the early 1800s, when Napoleon's propaganda machine needed anti-Russian slurs, one of those that really caught on was '*Grattez le russe et vous trouverez le tatare!*' (scratch a Russian and find a Tatar) (Halperin 1987: ix). This saying is not without substance, however. An example from everyday life may be the Russian taboo against shaking hands across thresholds. A ritual example of the lingering importance of hybridization today may be found in that key object of anthropological inquiry, burial rites. In addition to the standard funeral, Russians come together 40 days later. The religious explanation for this is to do with shadowing the Ascension; the deceased's loved ones congregate to ease the soul's passing to heaven. Note that this is a common practice amongst Muslims, but not so amongst other Christians. Yet another example, this time from a core area for International Relations (IR), namely diplomacy, concerns the restrictions imposed on movements by diplomats which were in evidence continuously from the Mongol period until the fall of the Soviet Union, a shadow of which remains even today. Remnants of the way things were done during Mongol rule are still readily observable in Russian life.

Conclusion

Rus' should be categorized as a suzerain system of polities centred on Kiev, rather than as a single polity. The polities were lineages led by princes. Neighbouring powers, including steppe-dwelling peoples, were brought into the fight between lineages on a regular basis. Once the Mongols destroyed Kiev in 1240 and established a new layer of Mongol overlordship, this loose suzerain system of lineage-based polities characterized by a high level of conflict and open lines to allies from the adjacent steppe became part of an imperial structure – that of the Golden Horde. For some decades afterwards, the Khipchak Khanate was itself still part of an imperial structure. The Khipchak Khanate ruled Rus' according to standard operational Mongol procedures. At the beginning of the fourteenth century, two lineages, now thoroughly territorialized in the cities of Moscow and Tver', fought for predominance amongst the Rus'. Moscow owed its victory to the superior way in which they had played the alliance game vis-à-vis the Mongols compared to other Rus' polities. From the 1370s on, in order to underline how Moscow was changing the suzerain system of Rus' lineages into a polity centred on Moscow, it is customary to refer to this polity as Muscovy. Muscovy was still subservient to the Khipchak Khanate, and would remain so for another 100 years, until the Golden Horde fell apart in the first decade of the 1500s.

As Mongol suzerainty waned, relations with steppe polities nonetheless continued. When, in the early decades of the fifteenth century, rival Mongol khans were not able to maintain order amongst the local Tatar princes of the Dnieper steppes, some of these formed semi-independent detachments that became known as Cossacks. Vitautas of Lithuania hired some of them to man the steppe frontier. Slavs who were similarly employed also came to be known as Cossacks (Vernadsky 1953: 289). When, in the 1440s, Moscow decided to resettle the Tatars who had joined its grand duke's service, 'the best solution seemed to be to establish a network of advance posts along the southern border of Russia, close enough to the Tatar-controlled steppes so that Russia's military leaders could both watch the movements of the Tatars and repulse them when they came' (Vernadsky 1953: 331, comp. 320). A former khan's son, Kasim, had a claim on a particular stretch of the frontier around the town of Gorodets, and so in 1452–53 Moscow created a separate polity for him there. Gorodets was renamed Kasimov upon Kasim's death in 1471, and went on to become a separate khanate which survived until 1681, as Muscovy's vassal and serving 'primarily as nomadic auxiliary troops' (Halperin 1987: 109). By then, Muscovy had annexed all the Khipchak Khanate's successor states.

As was demonstrated most recently by the key role played by Russian federal subjects such as Tatarstan and Bashkortostan during the dissolution of the Soviet Union (Neumann 1999: 183–206), successors of those successor states are still a distinct presence in Russian politics. Through the sixteenth and seventeenth centuries, it was fashionable for Russian aristocratic families to sport their Mongol connections. After Peter the Great's reforms, with the fading of the Crimean Khanate, which was the Khipchak Khanate's principal successor state in cultural terms, and with Russia's increasing Siberian expansion, the Tatar experience took on a more and more subterranean role in Russian historiography. The role of the Tatar and also of other identities such as the Kalmyk (successors of a specific Mongolian tribe) to present Russian national identity awaits further study, but there has certainly been ample hybridization. Tatars, Kalmyks, Bashirs and other groups whose collective memory is tied up with having been part of the Khipchak Khanate remain liminal to Russian identity.

The Mongol connection and the hybrid character of the polity of Muscovy coloured Russian entry into the European states system. Muscovy itself chose to seek recognition from the continental European powers, with which connections increased steadily in the fifteenth century and after the fall of the Mongol Khipchak Khanate, as successors to that Khanate. The bid for recognition was presented by dint of a number of practices that were taken directly from the Mongols. Continental European powers were therefore warranted in seeing Muscovy as a partly Asian polity.

It should also be clear, however, that the political logic of what was going on in the North, between Scandinavians, Lithuanians, Poles, Germans, the Rus', the Khipchaks, etc., was one where confession had importance, but not necessarily overwhelming importance. It is simply not the case that an overarching polity, be that Christendom or its successor Europe, stood against other polities. Neither is it the case that the continental European powers imposed a ready-made system of interaction on Muscovy (or on other Northerners, for that matter). It is very hard to identify a clear geographical, social or political boundary between Europe and non-Europe in the period under discussion here. Rus' polities before the Mongol intervention had interaction with steppe polities as one of their defining traits from the very beginnings in the eighth century.

I have argued that the area called Rus' which the Mongol forces subdued at the end of the 1230s was a loose suzerain system of lineage-based polities characterized by a high level of conflict and open lines to allies from the adjacent steppe. I have also argued that the establishment of a Mongol imperial order centred on the Golden Horde and lasting for around 250 years meant that when Muscovy emerged as the Golden Horde's self-acknowledged successor polity, it was as a hybrid polity with state institutions and diplomatic practices that bore deep marks of its steppe heritage. Furthermore, Muscovy's emergence came as a result of, among other things, a century of alliance politics where the principal actors were the Khipchak Khanate, Muscovy and Lithuania (Lithuania/Poland). It is an indictment of IR as a discipline that its focus has tended to be so narrow that not even inter-polity relations in the immediate vicinity of Europe such as these have been deemed worthy of much attention.

There are a couple of lessons to be drawn here. First, the editors are right to point out in their Introduction that we should not think of the past as being culturally homogenous. From the 220s BC, steppe relations were about building up multiethnic empires, which sustained themselves, among other things, by attacking sedentary polities to the south: China, Persia, the Byzantine Empire, etc. If, as in the fourth century, a steppe empire had no luck in China, it could regroup and attempt a devastating attack on the Roman Empire instead. As a result of all the ensuing hybridization, of which Russia is a key example, there simply is little or no cultural ground on which to found a division of the world into discrete civilizations (cf. Bowden 2009).

A second lesson to be drawn concerns the status of nomadism in world history. Barry Hindess has noted about present-day migration discourse how:

> The assumption here is that, even if they move around within it, people will normally be settled in the society to which they belong … In fact, the historical record suggests a different story; namely, that large-scale population movement is as normal a feature of the human condition as is long-term territorial settlement … Nevertheless, the system of territorial states and the techniques of population management developed within it have turned the movement of people around the world into an exceptional activity, something that can and should be regulated by the states whose borders they threaten to cross.
>
> (Hindess 2000: 1494)

This is certainly so. All European peoples (with the possible exception of the Basques) hail from the steppes, and it took millennia before there was any meaningful distinction to be made between the two. Let us not forget that the first occurrence of the concept of Europe in mediaeval history hails from the crowning of Charles the Great in the year 800. He crowned himself emperor, amongst other things, to celebrate his victory over the Avars, a steppe people. In a very real sense, the Russian experience with the steppe is an historical coda of the European experience with the steppe.

A third point goes to the core of this volume. In a European comparative perspective, with the exception of the Balkans, Russian experiences with non-Europeans were particularly long lasting, and they included 250 years of suzerainty. In a global comparative perspective, if we juxtapose Russian-Mongol relations with the European-non-European relations that were to follow in the early modern period and which are the topic of the other chapters in this book, the similarities are overwhelming. Even more interesting is the fact that Russians, being Christian notwithstanding, in many ways came to be Other-ed in the very same way as (other?) non-Europeans in that period. I have discussed this topic at some length elsewhere (Neumann 1996, 1999). Suffice it to point out that the roots of critical development theory are to be found in the writings of a Russian Jew, namely Leo Trotsky. It was his reading of Russia's development as 'combined and uneven' in a world dominated by (Western) Europeans that formed the template on which intellectuals in other parts of the world began to theorize development. To the degree that there is a line to be drawn from critical development theory to post-colonial scholarship, and to the extent that this book embodies the former, the structural parallels between Russo-European relations on the one hand, and Chinese-European, African-European, Latin American-European relations, etc., on the other have their counterparts in the knowledge production about these relations in Russia on the one hand, and other countries such as China, India, Turkey, Persia, Algeria, Brazil, etc., on the other.

I noted above a propagandist put-down from the court of Napoleon: 'scratch a Russian and find a Tatar.' There is nothing empirically wrong in this statement. Russian culture is a thoroughly hybridized phenomenon. What is wrong with this slogan is the modernist preconditions that lend it its negative propagandistic force, namely that hybridization is bad. That value judgement was not predominant in the world before the arrival of the anarchical society. It is likely to be buried together with the modernity of which it was such a characteristic part.

Notes

1 The ruler of Muscovy, who had taken the title of tsar in 1547, annexed the successor polity of the Khanate of Kazan' in 1552 and the Khanate of Astrakhan in 1556.

2 Beyond the Khitans, there is an uninterrupted tradition of steppe empires reaching back for at least 1,500 years. From the perspective of their neighbours to the south, the rise of the Mongol empire was a working accident: 'There was a standard imperial Chinese policy for dealing with them. They would be carefully watched,

and if one nomadic chief seemed to be gaining power and influence at the expense of others, Chinese subsidies, recognition and titles would be offered to one of his rivals, who would be encouraged to cut the upstart down to size. Should the new protégé in his turn seem to be becoming dangerously powerful, the process would be repeated' (Morgan 1986: 35).

3 The break came in the late 1250s, when the Khipchak Khan Berke broke ranks. The Great Khan in Karakorum had assigned the Caucasus to Batu's Khipchak Khanate. The Great Möngke reversed this decision, giving it to the rivalling Mongol polity of the Persia-based Il-khans instead. The Caucasus remained a bone of contention between the Khipchak Khanate and the Il-khans (and also to their successors, the first of whom was Timur-lenk) and *mutatis mutandis* down to the present era. When Möngke died, the struggle over the Caucasus became the main factor in determining the Khipchak Khanate's and the Il-khan's positioning in the succession struggle. The Il-khan candidate (Hülagü) won. The leader of the Khipchak Khanate (Berke) answered by taking a step unprecedented in Mongol imperial history – namely to forge an alliance with Mamluk Egypt against his fellow Mongols, the Il-khans. The 1261 alliance was followed up by a commercial treaty, which opened up for trade that proved lucrative to both sides (basically slaves for luxury goods). This trade was of utmost importance for the Golden Horde until, in 1354, the Ottoman Turks took over control of the Dardanelles from the Byzantines. The Ottoman Turks effectively put an end to the Golden Horde's Egyptian trade.

4 He was followed by his brother Qubilai (Kublai Khan, 1260–94). Qubilai concentrated on China, and was not much of a presence in other parts of what was now increasingly the former Mongol empire.

5 A correspondence is often assumed between the four sons and the subsequent Mongol-led polities in China, Persia, Central Asia and Russia, but as pointed out by Jackson 1999, this is too neat.

6 The Alans were a Farsi-speaking people (and so by the lights of the day arguably further removed from the Mongols than Turkic-speaking peoples like the Khipchaks). Eventually, a large number of them settled in Khanbaliq (now Beijing), where they were converted to Christianity by Archbishop John of Montecorvino. They became a mainstay of the Mongol army. Kagan Toghon Temür sent an embassy to the Pope in 1338, asking the Pope to send a new pastor, as well as for his blessing. De Rachewiltz (1971: 188) sees the key reason for this as being the kagan's 'desire to please the military chiefs on whom depended the security of the state and the emperor's own safety'.

7 Soviet historians like Bartol'd have suggested that Batu was co-ruler, but Allsen (1987: 54–59) and others have convincingly refuted the argument.

8 For example, in 1256–57, Prince Gleb Vasil'kovich of Rostov journeyed to Karakorum, and returned with a wife, a Mongol princess (Allsen 1987: 183–84).

9 'For example, Cyril, the Metropolitan of Kiev, who at first supported the anti-Mongol princes of Galicia and Volynia, in the end (1252) threw his considerable weight behind Alexander Nevsky, the prince of Novgorod and champion of accommodation' (Allsen 1987: 122).

10 Since one of the points I am trying to make is that Europe's eastern frontier is a hybrid, it should be pointed out that the boundary just drawn is also in need of de-differentiation. For example, Mongol power resulted in 'a revival of the old steppe traditions at the court of Hungary' in the latter half of the thirteenth century (Vernadsky 1953: 180–81).

11 The Nogay, named after the Mongol Nogay Khan, based in the Caucasus around present-day Kalmykia and harbouring a number of Khipchaks, were at loggerheads with the rest of the Golden Horde in the 1290s, and established themselves as a khanate in 1319. They 'built a power base in the Crimea and the Balkans and

contested with the khans of the lower Volga for control of the Golden Horde' (Halperin 1987: 18).

12 The two other cities to be ruled by grand dukes, Nizjniy Novgodor and Riazan, came up short on both counts.

13 Vernadsky (1953: 167), whose major thesis throughout his multi-volume history is the rise of Russian nationhood (preceded by a 'federal' Kievan state period), nonetheless stresses how Alexander's brothers' and sons' failing to settle in Vladimir upon becoming its grand duke constituted 'the temporary victory of the apanage (*udel*) principle over that of the nation state'.

14 From the 1250s onwards Galicia and Kiev came ever closer to the Lithuanian kingdom, and were eventually absorbed by it. The Russian aristocracy asserted themselves strongly, however, to the point that a variant of Russian (White Russian) became the kingdom's official language. A number of nobles eventually gravitated back to Muscovy (cf. Backus 1957).

15 Tver' did not give up, though. In 1382, it allied with Khan Tokhtamesh of the Golden Horde against Moscow.

16 The fifteenth-century political cabal in these parts turned on a familiar alliance pattern. In the east the Golden Horde strove to hang onto its suzerainty in Russian lands, which were increasingly dominated by Muscovy. In the south the Crimean Khanate tried to stem the increasing influence of the Russians on the Golden Horde. In the west the Lithuanians tried to encroach on Russian lands. Logically, a basic alliance pattern emerged whereby the Golden Horde and Muscovy paired up against Lithuania and the Crimean Khanate. Once the Golden Horde unravelled, there was elite integration. During Russia's Time of Troubles, the Tatar aristocracy rallied to the Russian cause against the Poles. The Tatar aristocracy was placed side by side with the Russian one in 1784 (Spuler 1965a: 91). The Crimean Tatars had a rather different end of it. Their final raid on Moscow, in 1571, ended with them actually being able to force the Muscovites to pay, but that tribute never seems to have been paid. Once the hetman Bogdan Khielnitski transferred his loyalties from Poland to Russia in 1654, the Poles and the Crimean Tatars once again made common cause against Russia. The Crimean Tatars remained a potential ally for Russia's opponents until they were incorporated into Russia after the Russian victory over the Ottomans in 1774.

17 This is not to say that the relationship between Muscovy and this second most long-lived of the Golden Horde's successor states was not volatile. The Crimean Tatars burnt Moscow to the ground as late as in 1571. That, however, did not keep certain boyars in Muscovy from considering the Crimean khan as a possible ruler (cf. Ostrowski 2000).

18 Janet Abu-Lughod's interesting attempt to theorize the world system before European hegemony is marred by her specious readings of these reports. Although she herself notes that their use of imagery is of the same kind (and frequently even parading the same specific ideas about monsters and strange humanoids) as contemporary Chinese texts about western lands, she does not hesitate to heap scorn on leading European scholars of the period like William of Rubruck (Abu-Lughod 1989: 162; comp. de Rubruquis 1990).

19 As late as the seventeenth century, the émigré Muscovite bureaucrat Gregorii Kotoshikin explained that the ruler of Muscovy was a *tsar* by virtue of Ivan IV's conquest of Kazan (Halperin 1987: 100).

2 Europe, Islam and *Pax Ottomana*, 1453–1774

Ayla Göl

Introduction

Historical diplomatic relations between European powers and the Ottoman Empire (1299–1923) are usually regarded as the most enduring and equal of Europe's relations with the non-European world. Though the Ottoman Empire was historically and geopolitically embedded in the 'diplomatic system of Europe' through warfare, trade and bilateral agreements (Mansel 2010: 15), it was never regarded as part of Europe and did not know its place in the hierarchy of European powers until it was formally admitted as the 'sick man of Europe' in 1856 (Adanır 2005). In the narrative of the English School on the expansion of international society, the Ottoman Empire was the first Islamic entity to be admitted to international society in 1856 (Bull and Watson 1984b). The 'inclusion' of the Ottoman Empire was followed by Japan and China in the late nineteenth century, as non-Christian and non-European civilizations gave the expanding European international society its multicultural and 'universal' character (Naff 1984; Zhang 1991; Yurdusev 2003; Suzuki 2005). However, the idea that the West was keenly engaged in establishing relations with other civilizations has usually been presented from the perspectives of European powers. As Paul Keal (2000: 64) rightly points out, the inclusion of non-Christian and non-European states is 'a vital but often neglected part of the story of international society'.

While 'high imperialism' of the late nineteenth century integrated different cultures, civilizations and religions into a hierarchical international society, this process was arguably the consequence of an unprecedented expansion of European imperialism and domination to the other parts of the globe in the nineteenth century. One intellectual legacy of this process is that it produced a *Eurocentric model of the world* that put the dominant West at the centre and thereby tacitly relegated the rest of the world to subordinate positions (Hobson 2004). As the editors of this volume argue in the Introduction, an immanent critique of Eurocentrism is necessary not only to understand the parallel historical experiences of non-European societies in other regions of the world, but also to explain the existence of a plural international order.

Such an exercise is also useful in overcoming conventional understandings of Ottoman history, which has reduced the complexity of the Ottoman Empire's political, territorial, military and economic structures into one of an *Islamic* order (cf. Heywood 2002: 431). Until the emergence of a revisionist group of Ottoman historians in the mid-1980s who published in English (Imber 2002; Heywood 2002; Finkel 2005; Mansel 2010), the approach to studying Ottoman history was a product of 'myths and stereotypes' that portrayed the Ottoman Empire as a social 'outsider' (Heywood 2002: 431), a view also found in the narratives of the expansion of international society (Bull and Watson 1984a). However, as Finkel (2005: xiv) describes it, the 'black hole' that is Ottoman history stems mostly from the West's 'old narrative' of the Ottoman Empire, which is an integral part of producing orientalist discourses of the Islamic past for many centuries (Halliday 2009: 1). In the twenty-first century, it is a matter of urgency to remove the 'iron curtain' of misunderstanding between the West and Muslims in general and the status of the Ottoman Empire in the international order, in particular before the arrival of anarchical international society. Against the backdrop of these main ontological assumptions, this chapter aims to contribute to the critiques of the Eurocentric approach of the English School by examining the 'rise and dominance' of an Ottoman-centred regional order at the crossroads of the Balkans in Europe, the Middle East and North Africa, with specific reference to 1453–1774.

A critical examination of the Ottoman history during this period has the added value of making empirical contributions to theoretical arguments on three issues. First, the rise of Ottoman power in the Balkans between 1453 (the date of the Ottoman capture of Constantinople, present-day Istanbul)[1] and 1774 (the symbolic date of the Ottoman decline after the Treaty of Kutchuk Qainarji – *Küçük Kaynarca*) indicates that not only had the Europeans been in contact with the Ottomans as a non-European actor much earlier than the nineteenth century, but also Muslim Turks were in a powerful position after the capture of Constantinople to dictate their norms and rules on the basis of European *weakness* in their relations with Western powers. Second, the Ottoman dominance in regional politics between 1453 and 1683, the so-called '*Pax Ottomana*' (the Ottoman peace) in Anatolia, the Balkans, the Middle East, North Africa and the Caucasus, contests the assumption that international relations between Europeans and non-Europeans is characterized by the domination of Western powers (Çiçek 2001; Ortaylı 2007: 7). Third, the Ottoman Empire's diplomatic and economic relations as well as warfare with major European powers (Venice, France, Austria-Habsburg and England) in the West and non-European powers (Iran and Russia) in the East help us to understand the existence of a plural international order before the arrival of the anarchical European international society in the mid-nineteenth century. It also shows that the Europeans were not the only actors in the diplomatic relations of the Ottoman Empire.

The chapter proceeds as follows. The first part of this chapter provides a critique of the expansion of international society as argued by the English

School of International Relations (IR), with particular reference to the place and status of the Ottoman Empire. In particular, it offers an alternative historical account of the rise of the Ottoman principality in Europe and the cultural hybridity of the Ottoman system, which straddled both Western and Eastern civilizations. The next section examines the Ottoman expansion towards Europe after the capture of Constantinople, which highlights the superiority of Ottomans and how the empire set up the regional order of the *Pax Ottomana* and imposed their rules and norms on European powers. Here, I focus on the dominance of the Ottoman-centred regional order – *Pax Ottomana* between 1453 and 1683 – and offer an empirical critique to Eurocentric narratives of IR, which frequently assume Western ascendancy. The last section highlights the end of *Pax Ottomana* and the retreat of the Ottoman Empire in the post-Westphalia international system in the eighteenth century. The chapter concludes by highlighting the 'added value' that the study of the Ottoman Empire offers for more recent debates over civilizational interactions and pluralism, as opposed to the longstanding claims within the English School that a common (Western-based) culture is necessary for establishing international order.

Rethinking the expansion of international society and the Ottoman Empire

The expansion of international society to the non-European world has become one of the dominant themes of the English School. The first important work examining the case of the Ottoman Empire was Thomas Naff's contribution to Bull and Watson's (1984a) edited volume on the expansion of international society.[2] As highlighted by Bull and Watson (1984c: 427), the Ottoman Empire was regarded as 'not only non-European in culture and race, but the historic enemy of the Christendom'. Neumann and Welsh (1991) subsequently introduced a new research agenda for scholars of international society by arguing that the Ottoman Empire was the 'other' in the definition of European identity. In an influential book, *The Evolution of International Society*, Adam Watson (1992: 312) maintained that social contracts in the form of regulatory machinery 'operated between the European *grande republique* and the Ottoman Empire since the mid-nineteenth century. Such contracts have been constantly revised and will continue to be revised'. It is clear that these scholars share the idea that the Ottomans were a 'social outsider' with limited 'societal' interactions with the European world.[3] However, this idea needs to be rethought by 'telling the story' of the empire from a non-Eurocentric perspective before the arrival of European international society in the nineteenth century.

The rise of the Ottoman Empire

The Ottoman Empire, as one of the 'three empires in the Mediterranean that survived to the twentieth century' (Ortaylı 2006: 7), dominated an area that not only included most of the territories of the eastern Roman Empire, but also large portions of the Northern Balkans in Europe and the north Black Sea coast at its zenith in the sixteenth and seventeenth centuries. Although most of the frontier regions in the Balkans, the Caucasus and the Near East experienced continuous warfare, most of the other provinces in the Anatolian hinterland enjoyed a long period of peace and prosperity under the rule of the Ottoman military and bureaucracy (Uyar and Erickson 2009: 281).

The rise of the Ottoman principality (*Osmanlı Beyliği*) (the word *Osmanlı* was corrupted to 'Ottoman' in English) from a small tribal polity to one of the most powerful Muslim empires on the margins of Europe across three continents was one of the most remarkable events of world history.[4] Even when the decline of the empire was first described as 'the Eastern Question' in European history after the defeat of Ottomans against the Russians in 1774, and then as 'the sick man of Europe' later in the mid-nineteenth century, the empire's rule still stretched over three continents: Asia, Europe and Africa. Geopolitically, it included the parts of the Balkans, the entire Near East (present-day Turkey and all the Arab states from Iraq to Libya), and North Africa (modern Tunisia and Algeria still had at least some nominal connection with Constantinople). This would continue until its ultimate disintegration after the First World War (Albrecht-Carrie 1958: 41; Quataert 2000: 5).

Historically, it was in 1352 when the Ottoman Turks under the leadership of Orhan Bey brought the Ottoman frontier to Constantinople, securing a foothold for the Ottomans on the European continent for the first time. In the fourteenth century, the Ottoman principality expanded into Europe under ambitious sultans such as Murad I (1326–89) and Beyazid I (1360–1403). In 1389, 'Serbia had first become an Ottoman vassal state after the battle of Kosovo' (Finkel 2005: 59). In 1396, the Ottomans won the Battle of Nicopolis (*Niğbolu savaşı*) against an allied force from Hungary, France, Venice and some other smaller contingents in Europe, which was widely considered as the last large-scale crusade of Europe's Middle Ages. These historical events are particularly important not only to emphasize the advance of the victorious Ottomans into the Balkans as a gate to Europe, but also to highlight the *weakness* of European power.

Before the capture of Constantinople, the Ottoman policies towards Western powers were not only characterized by warfare but also by the establishment of diplomatic relations and alliances in the Mediterranean. The Ottoman Sultan Bayezıd signed a trade agreement with Venice in 1388, and utilized it for strategic purposes to prevent Venice's involvement at the battle of Kosovo when the Ottomans attacked and established rule in Serbia as a vassal state. By maintaining good relations with the Turks, in return, Venice protected its commercial interests in the Ottoman dominions and the Black Sea region. Based on mutual commercial interests, Ottoman–Venetian relations remained friendly until the war between the two states from 1423 to 1430 (İnalcık 2009: 109; Shaw 1976: 47).

In the fifteenth century, the Ottoman dynasty under the rule of Murad II (1421–51) undertook a period of great expansion both in the Balkans (such as

Thessaloniki, Macedonia and Kosovo) and in Anatolia (such as Ankara and Konya) (Uyar and Erickson 2009: 1; Ortaylı 2007; İnalcık 2010: 60). In short, after the period of the establishment of the Ottoman principality (*kuruluş devri*) between 1299 and 1452, the age of the Ottoman Empire (*yükseliş devri*) between 1453 and 1683, *Pax Ottomana*, is generally regarded as the crucial era that transformed the empire into a world actor (Kunt 1995a: 23–30, 78–82, 120). The Ottomans continued to grow and involve themselves increasingly in European diplomacy at the turn of the sixteenth century. Under Beyazid II's rule (1481–1512), the external affairs of the empire were shaped by the military campaigns in the Balkans, the Mediterranean and Anatolia that set the stage for the Ottoman-centred regional order. The Ottoman–Venetian war (1499–1502) was particularly significant for the beginning of Ottoman dominance in the Mediterranean, which confirmed the Sublime Porte's status of a 'sea power' in the international arena (İnalcık 2010: 132–34).

At the height of their power, the Ottomans shared world history with powerful European states such as Habsburg Spain, Elizabethan England and the Holy Roman Empire, as well as Valois France and the Dutch Republic (Quataert 2000: 3). During this period, as Ortaylı (2007: 13) argues, '*Osmanlı*' was no longer the name of a tribal principality but an 'imperial identity' that indicated until 1683 the superiority of the Ottoman Empire over their Europeans counterparts and the rise of Ottoman agency in world history. After this date, the empire was in decline (*çöküş devri*) until 1922. Nevertheless, the Ottoman Turks continued to be perceived as a 'serious threat' to European collective security and identity until its complete disintegration and transition to a modern Turkish nation-state in 1923.

In *realpolitik* terms, between the fifteenth and eighteenth centuries, the Ottoman Empire was a de facto significant European power, controlling between a quarter and a third of the continent (Rich 1999: 443). As Watson (1992: 116) argues, the Islamic 'Turkish Empire that survived into the twentieth century' had the most remarkable impact on the 'European system and the evolution of international society'. The fall of Constantinople in 1453 was a crucial event that marked the rise of Ottomans and highlighted European *weakness*. It elevated the small tribal Anatolian principality into an imperial state by consolidating the expansion of Ottoman power into Europe and their military superiority over European states. Among other civilizations, Naff (1984: 145) states, the Ottoman Empire played a decisive role in two important movements in Europe: 'the Protestant Reformation and the Catholic Counter-Reformation, which together with the Renaissance transformed European society and led ultimately to the creation of an international society in Europe's image'. As Finkel (2005: 58) highlights, the 'fall of Constantinople to the Ottomans was a matter of horror for the Christian West, which feared an ever more aggressive policy of conquest'. From the European perspective, the Turkish attack reinforced the image of the 'dangerous Ottoman-Turk' and justified the defence of the Christian faith and identity (Neumann 1999: 45). Hence, the capture of Constantinople by the Muslim Ottoman Turks represented more than just the material loss of territory for the Europeans. It created the image of the 'barbarian Muslim Turk' and a real ideational threat to European security and Christian collective identity with reference to the 'consciousness of the Turk' (Coles 1968: 145). In particular, it meant that the Christian communities and churches of Asia Minor – Anatolia (*Anadolu*), which was considered part of the Near Eastern origins of 'modern Western civilization' – were no longer under 'Christendom' but Islamic Ottoman rule (Mann 2005: 31). Unsurprisingly, the fall of Constantinople in 1453 was regarded as a 'world-historical accident' (Mann 2005: 508) by European rulers and observers. In contrast, Turks celebrated this event and Ottoman historians considered it the beginning of a 'new era' (*Yeni Çağ*) in their history, signalling their superiority to the Christians (Kunt 1995a: 80–88).

Governing the Ottoman Empire

While the fall of Constantinople acted as a 'homogenizing' force for Europeans, it also had unprecedented consequences for Ottoman domestic and foreign policies, in that they had to enter into much denser relations with both their Eastern and Western neighbours. Domestically, a new social system emerged after the conquest of Constantinople, where the weak Byzantine state's practice of granting a series of economic privileges and extraterritorial rights to the Genoese, the Venetians and other Italian settlements in the fifteenth century was preserved and continued by the Ottoman state (Karpat 1973; Shaw 1976). Mehmed II (1444–46, 1451–81) kept some of these Italian colonies' special privileges in order to continue trade relations that had laid the foundations of the *millet* system, which was initiated to organize the remaining different religious and cultural groups. The *millets* were essentially religious communities with cultural autonomy. The leaders of *millets* were not 'considered alien to the system but an organic part of the bureaucratic apparatus. They maintained order and eventually collected taxes and fines' (Karpat 1973: 32–33). Hence, through the *millet* system, the peasant masses were linked to the central government, leaving little room for the use of religion by local leaders in the Balkans and Anatolia.

As argued by Ortaylı (2007: 11–12), the *millet* system was the basis of *Pax Ottomana*, which meant the harmonious existence of various *millets* under the Ottoman rule. *Pax Ottomana* was based on two principles: 'equality' between Muslims and non-Muslims (Hellenic-Christians and Jews); and 'peace' under the Ottoman governance. Similarly, it is argued here that the Ottoman Empire's dealings with its European neighbours shared common features with the *millet* system, in that religious differences were recognized and tolerated in order to establish regional peace and stability. In their foreign relations, the sultans did not demand cultural/civilizational assimilation as a prerequisite for diplomatic relations, as the Europeans would do later in the nineteenth century. Such an unorthodox interpretation of Ottoman history leads us to rethink the agency of Ottoman Empire in European international relations.

Reinstating the agency of the Ottoman Empire in European affairs

The Ottoman Empire became significantly involved in the international relations of Europe through warfare, diplomacy and trade relations, as well as its Islamic mission as the successor to the great Muslim caliphates of the past. However, as Naff (1984: 143, emphasis added) emphasizes, the 'logical conclusion ought to be that the Ottoman Empire was, empirically, a European state. The *paradox* is that it was not'. Indeed, the Ottoman Empire was never regarded as part of Europe until the mid-nineteenth century despite the fact that it still had large territories in the Balkans and was involved in European political and military politics. Nevertheless, it had to be taken into account as a major power by all the European states as long as it was militarily powerful and perceived as a threat to international order.

The paradox was deepened by the policies of European powers to include the Ottomans into 'systemic' relations but exclude them from 'societal' relations. The house of Osman had a different socio-political organization – its *millet* system – and a different religion and identity to the European states (Ortaylı 2006: 87–89). For Europeans, the fact that the Ottoman Empire had different principles of existence, institutions, norms and values from those of Europe made it a social 'outsider' as well as an historical enemy. 'In the world of the European mind, the Ottomans were alternately terrible, savage and "unspeakable" and at the same time' it was pictured as a militarist, authoritarian and despotic oriental empire, which was not recognized as a European state until 1856 (Quataert 2000: 7).

However, there is another side to the coin as well. As stated earlier, the story of the 'inclusion' of the Ottoman Empire needs to be told from a new perspective to challenge the prevailing Eurocentric character of international society. In contrast to existing accounts that imply that the Ottoman Empire was 'excluded' from the international order, it could be argued that the Ottoman Empire did not want to be a part of European international society or system during its climax. Arguably, the Ottoman sultans considered themselves superior to their Western counterparts and excluded the empire from the European society of states, rather than being 'excluded' by Europeans. As one of the most significant symbols of diplomatic relations, the history of exchanging ambassadors between the Ottomans and European powers support this argument. Interestingly, the Sublime Porte accepted the first permanent foreign diplomat, French Ambassador Jean de La Forêt, in 1535 and later English Ambassador Sir William Harbourne in 1583 and other European ambassadors to Constantinople at certain times for limited periods. Permanent Ottoman ambassadors were not sent to Western states until the eighteenth century (Shaw 1976: 181–82; Yurdusev 2009: 77).

Furthermore, our preoccupation with European international relations has also blinded us to the simple fact that the Europeans were not the only actors in Ottoman foreign affairs and the Ottomans also engaged in both Western and Eastern relations. In the early sixteenth century, once military advance and peace was secured on the European front, the Ottoman Empire would expand towards the Anatolian front in the east. This took place during Bayezid II's

(1481–1512) campaign against the Safavid Empire, which had taken advantage of Ottoman passivity in Eastern Anatolia and pushed their Shiite proselytizing efforts. In 1501, Shah Ismail (1501–24) came to power as the founder of the Safavid Empire in Iran and captured the city of Tabriz. Extending his influence in Iran and to Azerbaijan, the shah also sent many provocateurs to Anatolia in order to turn the Ottoman subjects against the empire. Afterwards, Shah Ismail launched a propaganda war against the Ottomans, which contributed to leading the empire into political chaos and a fight for the throne among young princes (*Şehzadeler*) (İnalcık 2010: 137).

In order to balance the threat from Safavids due to the increasing numbers of Shiite rebellions against the Sunni Ottomans in the eastern front, the Ottoman sultan accepted the Venetian request to sign a peace treaty in 1502 (İnalcık 2010: 135–36). After Yavuz Selim I (1512–20) dethroned his father with the support of Janissaries in 1512, he continued to expand the empire's eastern frontier and defeated Shah Ismail in the Battle of Chaldıran two years later. Sultan Selim I extended the eastern and southern frontiers of the empire between 1516 and 1518 by establishing Ottoman rule in Egypt, Jordan, northern Iraq, Syria, Lebanon, Palestine, Yemen and the Hejaz – the Holy cities of Mecca and Medina. In Ottoman history, taking vassalage of most Arab lands indicated the beginning of a new phase: the caliphate period, as will be explained later.

Hence, the 'rule and dominance' of the Ottomans in the Muslim world was another sign of its rise to great power status, which once again challenges the Eurocentric interpretation of international order, where the histories of non-European empires and their agencies in world history to construct regional orders have been consistently neglected. When the Ottomans defeated the Mamluk sultanate in 1517, they imprisoned the last Abbasid caliph, Al-Mutaakkil, in Cairo. Al-Mutaakkil was later sent to Constantinople, where he surrendered the caliphate to Selim I, which started a new era of a caliphate period in Ottoman history. Under the rule of Selim, the Ottoman territories spread to three times their former size, but it was his successor, Suleiman the Magnificent (1520–66), who further expanded upon Selim's conquests to make the empire a world power. The house of Osman thus not only became the 'protector' of the Muslim world against the 'infidel' Christian world but also established the Ottoman-centred regional order – *Pax Ottomana* – in the Balkans and the Middle East until the seventeenth century.

The age of *Pax Ottomana* in practice

The idea of Ottoman peace – *Pax Ottomana* – was shaped by the principles of equality and stability in the Balkans and the Arab Middle East. As indicated with reference to the *millet* system, the Sublime Porte recognized the equality between Muslims and non-Muslims – Christians and Jews – and, therefore, the Ottomans dealt with the Muslim and Christian worlds in a similar manner. While the Ottoman sultans prioritized state interests over cultural/civilizational ones when dealing with European powers, the Arab Middle East had become the centre of trade relations and enjoyed prosperity under the *Pax Ottomana*. Unlike its European counterparts, the Sublime Porte did not insist on the cultural homogeneity of the conquered provinces of the Ottoman Empire.

The principles of the *Pax Ottomana* in foreign affairs were supported by the policies of Suleiman the Magnificent in the sixteenth century: his *kanuns* (law) overruled Islamic law, which would not conflict with making a 'formal military alliance with an infidel' (Shaw 1976: 98). Whereas the Europeans took religious identity into account when forging alliances or entering into diplomatic agreements, the Ottomans under Suleiman the Magnificent were (perhaps surprisingly) less concerned about religious matters and civilizational differences. Instead, economic and military interests were a primary source of concern for the Ottomans. This tends to be ignored by Western scholars (Naff 1984: 148). Paradoxically, one can argue that although Sultan Suleiman had a world outlook less constricted by his religion, Ottoman–European relations were deep down partly influenced by *European* religious concerns, given that the Ottomans continued to be perceived by Europeans as the 'social outsider' because of their Islamic identity and culture.

Furthermore, the age of *Pax Ottomana* strengthens the arguments about the heterogeneity and multi-civilizational character of international orders. Yet, in the context of international relations, the Eurocentric view of the expansion of international society had led English School scholars to engage with its universal character. As Linklater and Suganami (2006: 74–75) pointed out, the institutions of international society had their origins in the European civilizations, but their expansion to the rest of the world brought a multicultural global international society. Bull and Watson (1984c: 429–33) previously argued that the European international society of the nineteenth century had cultural homogeneity, while the global international society of the late twentieth century became culturally heterogeneous. This interpretation of the expansion of international society creates an existential paradox for the English School: the universal and multi-civilizational character of international society challenges the hegemony of its common European civilizational origin (O'Hagan 2000: 115). The fact that the Ottoman-centred regional order in the Balkans, the Arab Middle East and North Africa allowed the Turks to produce a multicultural, 'hybrid' system of diplomacy supports the existence of the plurality of powers and cross-cultural engagement beyond the European self-definition and positioning at the centre of international order. One could argue that 'the eventual globalisation of the European system was not a foregone conclusion and it would theoretically have been possible for the Ottoman-centred system to have become the global norm with Europe on the periphery' (Bennison 2009: 57).

Europe on the margin of the **Pax Ottomana**

The rise of the Ottoman Empire as a major power between Europe and the Mediterranean in the sixteenth and seventeenth centuries is emphasized in this section to offer a critique of the two main claims of the English school: first, the Ottoman Empire was never regarded as a great power in Europe. Second, the Ottomans were 'excluded' from the international society because they did not share the European corpus of common interests, values, rules, laws, norms, customs, institutions and procedures until the mid-nineteenth century. These claims overlook the fact that the Sublime Porte reached its highest stage of dominance as an imperial power under *Kanuni* (the law giver) Sultan Suleiman's (Suleiman the Magnificent) reign and established the Ottoman-centred regional order. There were three important characteristics that gave the empire its supremacy: the Sublime Porte became an integral part of European diplomacy in the sixteenth century; a new system of capitulations was established, within which the Ottoman set the rules of trade and diplomatic relations with European powers; and the Ottoman policy of balancing between Western and Eastern powers was developed, which institutionalized the unique Ottoman diplomatic system that survived to the twentieth century.

First, the Sublime Porte became an integral part of European affairs and diplomacy as a major power. This is particularly important to challenge the key assumptions of the English School that the Ottoman Empire was not a 'great power' and was only formally 'included' in the (European) international society in 1856. The Ottoman supremacy can be traced back to the early sixteenth century: the capture of Belgrade in 1521 and Rhodes in 1522 was a manifestation of Suleiman's westward strategy to advance into central Europe while Europe was preoccupied with a war between the Habsburg Emperor Charles V and Francis I (1515–47) of France (Shaw 1976: 91). The Ottoman victory of 1526 at the Battle of Mohacs against Hungary allowed the Sublime Porte to advance further its interests in European affairs. The Ottomans were encouraged by the general atmosphere in Europe when the Habsburgs, the pope and England formed anti-French alliances that divided Christian unity. When Francis I appealed to Suleiman for help to restore the balance of power, the Ottoman sultan provided military and economic assistance to France against the Habsburgs. However, the Ottoman–French alignment was short-lived, and Francis I and Charles V ended their conflict under papal pressure to unite Europe against Islam (Shaw 1976: 99). The attempt to unite Europe against Muslims clearly did not prevent the first Ottoman siege of Vienna in 1529. As noted earlier, the Ottoman attack at the doors of Vienna strengthened the image of the 'barbarian Muslim Turk' as a real threat to European security and Christian collective identity since the fall of Constantinople.

The supremacy of the Ottoman power was not only limited to military affairs. It was also reflected in political and economic relations. For example, the Ottomans imposed their own rules in their interactions with European merchants. The 1536 French–Ottoman trade agreement was modelled on concessions given to Venice and Genoa earlier by the Sublime Porte. The Ottoman sultan was the only authority to decide the rules and norms of these concessions, which allowed the virtual freedom of French merchants from the Ottoman law (İnalcık 2010: 283). It was the *unilateral* decision of the Ottoman sultan that gave the right to the Europeans to travel and trade freely and to pay low customs duties on imports and exports within Ottoman territories. 'In essence, this made what came to be called the community of Franks in Constantinople a kind of *millet* and provided the mode for privileges subsequently bestowed on other European nations wishing to share the trade of the Levant' (Shaw 1976: 97). As mentioned with reference to the *millet* system earlier, cultural and religious differences were acknowledged and tolerated in order to establish trade and diplomatic relations, which laid the foundations of *Pax Ottomana*.

Second, and accordingly, a new system of capitulations was established, which not only highlights the supremacy of Ottomans but also the importance of economic interests in Ottoman-European relations. While analysing the expansion of the international society, the economic dimensions that shaped relations between the European and non-European worlds *prior* to the European ascendance have not usually been taken into consideration (but see Buzan 2004; Schouenborg 2013). The expansion of international trade to non-European markets is, therefore, another important dimension to add to the expansion of international society, which serves as another important critique of what is lacking in the existing English School scholarship. Early English School writers (see Naff 1984; Bull and Watson 1984a) largely disregard the importance of economic interests in the form of *capitulations* between Europeans and Ottomans. This results in ignoring the fact that as a result of the convergence of commercial interests with the Ottomans, Europe was – at least partially – an active participant in the Ottoman-centred regional order, even before the 'formal admittance' of the Sublime Porte into European international society in 1856.

It is important to note that the 'Capitulation system' that emerged during the sixteenth century was qualitatively different from its nineteenth-century counterpart, in that trading privileges were granted by the sultan alone as an imperial favour, and not under the threat of European military coercion (as was the case three centuries later). The trade agreements illustrated basic conceptual differences between the Ottomans and the Europeans. For the Ottoman Sultan, it was an *ahdname* (contract) that 'was not regarded as a formal alliance but rather a convenient instrument of policy, fashioned unilaterally against the Habsburgs' (Naff 1984: 148). Technically, there were no capitulations in the agreement. They were granted to France in 1569 only after confirmation by an *ahdname* of Sultan Selim II (1566–74). The unilateral character of these *ahdnames* reflected Suleiman the Magnificent's view of 'the inferiority of Christian Europe' and his belief that no European ruler *was equal* to the Ottoman sultan (Naff 1984: 146–48). These capitulations were granted by the Ottoman sultan *unilaterally* and they would become invalid at the sultan's death unless reconfirmed by his successor (İnalcık 2010: 283). The unilateral character of capitulations indicated the self-perception of the Ottoman sultan as superior to his European counterparts.

The privileges granted by the Ottoman Empire gave France commercial and political pre-eminence in the Middle East, lasting into modern times (Shaw 1976: 177). More importantly, the Ottoman–French trade agreement had some secret military and political terms that not only confirmed the existence of what could be called the beginnings of a 'capitulation system', but also indicated the first direct Ottoman–European interaction. In return, for the Ottomans, their alliances with France – usually against Venice and the Holy Roman Empire – was most probably aimed at decreasing the possibility of a European union in the form of another Crusade against the Turks, but it also contributed towards increasing the role of the Ottoman Empire in maintaining the European balance of power. This indicates the existence of enduring agreements between the Ottoman Empire and France, and challenges the claim that a (Western-based) common culture is necessary for establishing any meaningful relations between different entities.

The third characteristic that defined the Ottoman-centred regional order was that Suleiman the Magnificent developed a new policy of balancing Western and Eastern powers: the possibility of a two-front war based on the Holy Roman Empire and Safavid alliance was perceived as a real threat by the Ottoman sultan, and therefore the Sublime Porte did not wage war against the one without ensuring peace on the other front. Hence, the Sublime Porte did not shape its foreign affairs based on cultural or religious differences but strategic considerations. Moreover, the Ottoman Empire's sheer size meant that it was compelled to engage in diplomacy between two very different cultural entities simultaneously: that between Western Christendom and the Islamic world. The diplomatic practices that emerged as a result were not necessarily Islamic in character, but instead based on the bestowing of trading privileges and peace treaties. They were able to deal flexibly with different religious/cultural polities that bordered the Ottoman Empire. Ottoman historians agree that instead of either imposing its own structures or imitating the Western or Eastern models, the Ottomans had a tendency to preserve and transform existing practices into systems of their own (Kafadar 1995; Uyar and Erickson 2009; Ortaylı 2006: 16). Consequently, the Europeans were thus just one among many other participants in the series of diplomatic relations that the Sublime Porte had established under *Pax Ottomana*.

Acceptance of Ottoman superiority

In the late sixteenth century, the Sublime Porte continued its involvement in world politics by engaging systematic wars against Europe, as mentioned earlier. The Ottoman policy under the rule of Sultan Murad III (1574–95) was to gain the support and friendship of England, which already had a more direct interest in Mediterranean affairs. While Ottoman naval power was beginning to decline as a result of a series of costly battles in 1569–71, England was expanding its naval power and international commercial interests, within which Anglo–Ottoman 'formal diplomatic and commercial relations began in 1583 when Murad III granted Queen Elizabeth I a treaty of peace and friendship' (Horniker 1942: 289). When Queen Elizabeth I addressed Sultan Murad III as the '*Great Turk*', the sultan responded by implying that the

Ottoman Empire was not only an ally of England but also the protector of the queen. A registry copy of the letter from Murad III to Elizabeth in 1580 informing her that the privileges had been granted, ended as follows (Skilliter 1977: 115–16):

> And you, for your part, shall be steadfast in your submission and obedience to our door of felicity, and never cease from continually submitting and imparting the items of news which have occurred in those parts and of which you have been informed.

Afterwards, two English ambassadors were sent to Constantinople in 1583 in order to obtain a capitulation agreement. Interestingly, we do not see much evidence of the notions of European superiority that were to colour Europe's relations with the non-Western world, and still do to this very day. Queen Elizabeth appears to have demonstrated a degree of deference towards the sultan's power, and the latter's non-Christian identity does not appear to have been a hindrance towards establishing commercial and diplomatic relations on Ottoman terms. Similarly, religious concerns do not seemed to have stopped Murad III from accepting emissaries from an 'infidel' country, provided they were prepared to accept notions of Turkish superiority. A new capitulations agreement was signed between the English and the Ottoman empire on 3 May 1590, giving England a similar position to that of France in the trade of the Near East (İnalcık 2010: 308). Although (apart from the visit of the Ottoman Ambassador Mustafa to England at the Court of James I in 1607) the incidence of Turkish emissaries visiting England was remarkably low, the regular visits of English merchants and diplomats to the Sublime Porte ensured the continuance of this inter-cultural exchange (Ellis 1824: 83–88). English merchants were allowed to travel and trade freely in Ottoman dominions, and formed the Levant Company, which established the English commercial interests in the Middle East that lasted into modern times (Shaw 1976: 182).

These historical facts can be interpreted as indications that Ottomans were not interested in taking part in the diplomatic system of Europe but saw themselves as superior and accepted foreign ambassadors *unilaterally*, on Ottoman terms. Crucially, the Europeans did not insist on reciprocal exchanges of diplomats (as they would do later in the nineteenth century), as the Ottoman Empire was reluctant to establish a permanent diplomatic presence in European capitals during this time. It is therefore arguable that they were tacitly accepting the Ottoman sultans' visions of foreign relations. As long as the Ottoman Empire remained sufficiently powerful, this political arrangement continued, and was capable of dealing with most diplomatic interactions that took place between the European powers and the Sublime Porte for more than two centuries.

It could, of course, be argued that the somewhat unilateral nature by which the Ottoman rulers imposed these arrangements made it difficult for true 'societal' relations to emerge between the two parties. Such views could also be reinforced by the fact that religious and cultural identities seem to matter a lot more to the Europeans when compared to the Ottomans, who (in similar fashion to their *millet* system) were quite prepared to deal with this matter more flexibly. For instance, when the Ottoman armada was ready at the Albanian coast to join France in the invasion of Italy in 1537, the Turks – who had regarded the Ottoman–French trade treaty of 1536 as a coalition of forces (Shaw 1976: 98) – were surprised that the French failed to keep their promises. For France and other European powers, the Ottoman–French treaty was a temporary ceasefire with the Ottomans, and was not regarded as 'a treaty or alliance' *(alliance ou société)* (İnalcık 2010: 158). This was probably the first lesson to be learned by the Ottoman sultan about the nature of European politics that 'infidel friends would abandon all agreements when it suited their interests in Europe to do so' (Shaw 1976: 98). The change of alliances indicated that it was the power of the pope and Christian ethos ('*la république chretienne*') that prevented France from forming a military alliance with a Muslim ruler against another Christian country (İnalcık 2010: 158).

While these historical facts serve as a powerful reminder not to overstate our case for the emergence of deep social relations between the Ottoman Empire and the European world, this does not mean that the construction of an international order was impossible or did not take place. Here, it is worth reminding ourselves that European unity against the 'Muslim other' at times only came about as a result of *papal pressure* (as noted above), and in its absence the Christian Europeans were more than capable of entering some form of diplomatic relations with the supposedly 'infidel' and 'dangerous' Muslim Turks. Furthermore, the distinction between 'systemic' and 'societal' relations as conceptualized by the English School is ultimately unhelpful: as Alan James (1993: 273) argues, 'any meaningful interactions between two polities is extremely difficult without some societal relations emerging at some point, and it is equally hard to establish some objective "tipping point" when "systemic" relations become "societal"'. The so-called 'inclusion' of the Ottomans into European relations strongly challenges the English School's theoretical debates and binary juxtaposition between international 'society and system' (James 1993), which simplify the complex socio-political and economic relations that took place across historical and religious-cultural lines. Neither does the Ottoman Empire's unilateral imposition of the rules of diplomacy vis-à-vis the Europeans, nor its initial reluctance to 'assimilate' itself into European rules of international engagement, suggest that the Turks and the Europeans were incapable of establishing any form of socio-political order between themselves.

The end of *Pax Ottomana*

The Ottoman Empire continued to dictate the terms of interaction vis-à-vis the Europeans, despite its alleged 'exclusion' from European international society. This would only change when its decline finally forced both the Europeans and the Ottomans to alter their stances. In the mid-seventeenth century, the Ottomans continued waging systematic wars against Europe. It is worth emphasizing that during Sultan Mehmed IV's reign of 39 years (1648–87), the empire's territory had reached the largest in its history, and the second siege of Vienna in 1683 became the symbol of the involvement of the Ottoman Empire in European affairs.

Nevertheless, the decline of the Sublime Porte was inevitable, due to internal crises and the political intrigues within the Ottoman palace. The defeat of Ottoman forces by the Austrian and Polish armies in Vienna signalled the end of Ottoman military superiority, which, to a certain extent, decreased the perception of the Ottoman threat to Christian collective identity and security. As Quataert (2000: 2) argues, 1683 marked the 'permanent reversal of power relations between the Ottoman and Habsburg empires', and after this date the 'Ottomans never threatened Central Europe again'. Mustafa II (1695–1703) continued to engage in military campaigns against the Habsburg monarchy, the Polish-Lithuanian union and the republic of Venice during his reign. After the defeat of the Ottomans by the victorious European powers, a two-month peace congress between the Ottoman Empire and the Holy League – including Austria, Poland, Venice and Russia – took place. The congress was concluded by the signature of the Treaty of Karlowitz – Karlofça – in 1699, ending the Austro–Ottoman war of 1683–97. It was the first treaty in which the Ottomans lost their territories and, therefore, it has been accepted as a landmark for the beginning of the Ottoman decline. Thereafter, although the Ottomans struggled to dominate the territories in the Balkans, they stayed in control of south-eastern Europe – including the modern-day states of Bulgaria, Serbia, Greece, Romania and others – for 200 more years (Quataert 2000: 2).

The impact of the international treaty of 1699 was rather paradoxical in Ottoman–European relations, in that while the Ottoman state was seen as in decline, it was given an equal status to take part in European affairs. It was in 1699, when Europeans gained self-confidence about their military superiority after the defeat of Ottoman forces, that they decided to invite 'the dangerous Ottoman-Turk' to participate in a European congress for the first time (Naff 1984: 150; Neumann 1999: 51). After this date, the survival of the Sublime Porte in the European state system dominated the agenda of European powers for an entirely different reason. The new rising power in the East was the Russian Empire. Historically, Ottoman–Russian rivalry was shaped by the dictates of geography, at the heart of which was a struggle for the control of the Turkish straits. This struggle caused the animosity that precipitated 13 Ottoman–Russian wars over the course of four centuries, the first in 1676 and the last in 1914 (Rubinstein 1982: 1–2). Consequently, the Sublime Porte continued its historically rooted 'policy of balance', which was based on securing a Western ally against the Russian threat in the East (Armaoğlu 1987: 43). Moreover, the new diplomacy was compatible with rules and principles of the European balance of power system. The European diplomatic relations evolved around the so-called 'Eastern question' over the next two centuries. As Lenczowski (1990: 32) argues, 'more concretely, the Eastern Question may be narrowed down to the manoeuvring of various European powers to prevent Russia from encroaching too much upon the integrity of the Ottoman Empire'.

Thereafter, the Ottomans retreated from European politics and closely involved themselves in eastern affairs against the Russian threat. Hence, the Europeans were thus just one among many other participants in the series of diplomatic relations that the Sublime Porte had established under its rule. For instance, when Russia took advantage of internal turmoil in Iran and attacked it in 1724, the Ottomans launched a military campaign to protect Iran. The Russo–Ottoman treaty was signed in Constantinople, which left Azerbaijan to the Ottomans and Dagestan to Russia. In internal politics, Ottoman history had been identified with cultural reforms – the Tulip Era (*Lale Devri*) as a time of literary, cultural and artistic improvement – under Ahmed III (1703–30), and the military reforms of Mahmud I's (1730–54) reign. During this period, the empire's external relations were shaped by continuing wars with Iran over the control of the Caucasus. During Sultan Mustafa III's (1757–74) rule, the Ottoman–Russian rivalry continued in the Near East. The Ottoman defeat by the Russians in the war of 1768–74 led the Sublime Porte to recognize the need for European allies to protect the integrity of the Empire once again. When the war was concluded with the Treaty of Kutchuk Qainarji (*Küçük Kaynarca*) in 1774, it became the most significant symbol of the Ottoman decline. According to this treaty, it was formally accepted that Russia would have a permanent ambassador in Constantinople and it would also have the right to protect all the Orthodox Christian members of the Ottoman Empire (Kurat 1990: 28).

The treaty of 1774 was also of great importance in the history of the Near East. Crimea gained its independence. For the first time, the Ottomans agreed to pay war compensation to another country. This set the tone of relations between Russia and the Ottoman Empire until 1914; it established the principle of foreign interference in the Ottoman Empire, and with the capture by Russia of some of the coastline of the Black Sea, it led to Russia's involvement with the straits, which was not removed until the twentieth century. The stark reality following this treaty was that the Ottoman Empire was in decline and could no longer be defended without the support of European allies. However, the Ottoman Sultan Abdul Hamid I (1774–89) turned the Treaty of Karlowitz into a diplomatic victory in the east by claiming the power of the Ottoman caliph as protector of Muslims in Russia (Kunt 1995b: 46–49). As discussed earlier, it was in 1774 that the title of the caliph was used for the first time and acknowledged as having political power outside Ottoman borders.

Paradoxically, the significance of the Ottoman caliph increased when the Ottoman borders were shrinking and European politics were being gradually 'secularized' after the establishment of the Westphalian system. 'The universality of the Ottoman caliphate became a fait accompli in the nineteenth century' as a consequence of the need of Muslim leaders to unite around one central institution in order to oppose Europeans (Karpat 2001: 49). Meanwhile, Britain and France established close relations with the Ottoman Empire as continuation of the earlier capitulations agreements throughout the late eighteenth and early nineteenth centuries; what was different, however, was that trade privileges were demanded by the European powers and awarded with reluctance from the increasingly feeble Sublime Porte. Selim III (1789–1807) was the first sultan who recognized Ottoman weakness and the importance of being part of the European states system. Choosing to stay out of European diplomatic institutions was becoming less of an option, and the Sublime Porte established permanent embassies in major European capitals. Hence, the first Ottoman Embassy was opened in London in 1793 as an indication of opening diplomatic channels and communication between the two empires and civilizations (Neumann 1999: 53; Yurdusev 2004: 5).

During the new phase of Ottoman-European relations, Selim III benefited from Anglo–French support to achieve his military reforms of the New Order (*Nizami Cedit*) period (Kürkçüoğlu 2004: 132). As discussed earlier, Britain and France established their commercial and political pre-eminence in the Middle East that would survive to modern times. It was not an historical coincidence that Britain became the major trading partner with the Anglo–Ottoman Convention of 1838 and the reforms in domestic politics and the system of taxation were promised by the Ottoman state in 1839 and 1856 (Quataert 1994: 764). The Treaty of Paris was signed in 1856 to end the Crimean War of 1853 between the Russians and the Ottomans, and had a distinctive meaning for Ottoman–European relations. On the one hand, it recognized the continual disintegration and decline of the empire; on the other, the Sublime Porte was 'admitted' to the European society of states (Ortaylı 1987: 90). It is important to emphasize that the signature of the Treaty of Paris in 1856 was closely related to this economic dimension and should not only be interpreted in political terms as the 'formal admittance' of the Ottoman Empire into the expanding European international society. When the Ottoman ruling elite decided to sign the Treaty, they could not anticipate its long-term socio-economic consequences. As Braudel (1984: 483) argues, after this date, Ottoman interactions with the industrialized, capitalist and expansionist Europe through trade and diplomacy would sound the death knell of its greatness as an Islamic empire.

The final and the longest century of the Ottoman Empire was the nineteenth century, which was acknowledged as a period of reform and revolution (Ortaylı 1987). In foreign affairs, the Ottoman Empire lost its status as a 'great power' in Europe when all its Balkan territories were stripped away by the Treaty of Berlin in 1878 (Quataert 2000: 2). Nevertheless, Ottoman rule remained in the Middle East and North Africa until the end of the First World War. In domestic politics, reforms were introduced to save the empire, but many agree that the imperial reforms accelerated the collapse of the empire and precipitated the Turkish Revolution of 1908 that led to the rise of Turkish nationalism and the establishment of the modern Turkish state in

1923. However, as Zhang (1991: 9) argues in the case of China, it is important to recognize that these reforms constituted what Bull described as 'domestic processes of political and social reform which "narrowed the differences between Asian and African political communities and the political community of the West, and which contributed to a process of convergence"'.

Concluding remarks

This chapter has forwarded an alternative interpretation of Ottoman history by outlining the 400-year prelude to its formal 'inclusion' in the (European) international society in 1856. The historical analysis of this chapter seeks to contribute to theoretical arguments by adding the value of the empirical study of the Islamic Ottoman Empire. Three general findings stand out: first, the chapter supports and advances the general critique of Eurocentrism in the English School approach to understanding the expansion of international society by emphasizing the necessity of a pluralist approach to the study of international relations that goes beyond the arrival and subsequent 'expansion of international society' in the mid-nineteenth century. Based on the critique of Eurocentric perspectives of this particular historical epoch, it argues that instead of being 'excluded' by the Europeans from the European society of states, historically, the powerful Ottoman Empire did not seek to be 'included' in the European international society or system. Rather, some European states participated in *Pax Ottomana* on the conditions and terms dictated by the Ottomans. Such an interpretation highlights the agency of the Ottoman Empire as an independent and dominant non-European actor in its own constructed international order in a certain historical period.

Second, the 'rise and dominance' of *Pax Ottomana* at the crossroads of the Balkans, the Middle East and North Africa between 1453 and 1683 shows not only how Europe was, in fact, at the margins of the Ottoman-centred regional order, but also highlights the existence of a plural international order, where the European international system/society was just one of many international/civilizational orders with which the Ottoman Empire interacted. Moreover, the supremacy of Ottoman rule was historically evident in three key events: the capture of Constantinople, the establishment of the caliphate period and the capitulations granted unilaterally by the Ottoman sultans. These events put the Ottomans in a powerful position to dictate their norms and rules on the basis of European *weakness.*

Hence, from the Ottoman Empire's perspective, it is erroneous to assume that the Sublime Porte was never a part of the 'European international order', as diplomatic relations between the two had continued almost uninterrupted for more than 200 years. Conventional analyses stress that only after the Ottoman Empire's decline and its ability to exercise its agency towards the Europeans diminished could it be considered to have become a 'participant' (albeit one whose position was often questioned) in European international society. This analysis is in line with Karl Polanyi's (1957: 8) argument that in 1856:

> … the integrity of the Ottoman Empire was declared essential to the equilibrium of Europe, and the Concert of Europe endeavoured to maintain that empire; after 1878, when its disintegration was deemed essential to that equilibrium, its dismemberment was provided for in a similarly orderly manner.

Of course, this perspective only serves to reinforce the view that the acceptance of European/Western norms and rules is 'the only game in town' when it comes to establishing an international order on the basis of shared values and culture. This, arguably, is the consequence of the Ottoman 'barbarian' image due to its Muslim identity and culture, as mentioned with reference to orientalist views of Islam. More importantly, it was a consequence of an unequal expansion of European imperialism and unprecedented domination over the other parts of the world in the nineteenth century. One can argue that while the expansion of (European) international society brought different cultures, civilizations and religions together into a hierarchical international society, it also produced binary oppositions by placing Europe at the centre and the rest of the world at the periphery. Such a hierarchical order implied the imposition of Western culture, norms and rules over the non-Western world and left no room to acknowledge the complexity and interdependence of interactions between Western and non-Western civilizations.

Third, by highlighting the relative lack of Ottoman concern towards civilizational/religious differences in their relations between *both* Western and Eastern civilizations, I hope to have demonstrated that 'coexistence and symbiosis' between Islamic and Christian civilizations was 'possible and probably more common' for centuries before the arrival of (European) international society in 1856 (Kafadar 1995: 19). This seems to indicate that international order need not be based on a 'common culture', but can, and has, at least in *Pax Ottomana*, had a multicultural and heterogeneous basis. The English School has often concerned itself with the question of the need for a common civilizational/cultural basis for the establishment of international order. Such debates have at times been undergirded (albeit implicitly) by the belief that the acceptance and internalization of 'Western values' will bring about stability and moral progress in an anarchical international realm, but does international politics have to be established on the basis of Western cultural homogeneity? As shown in other chapters in this volume, we have seen that non-European polities and the Europeans were at times perfectly capable of entering into diplomatic relations in the absence of cultural domination, and in some cases the non-Europeans were able to impose *their* value systems on the latter. The Ottoman case, however, demonstrates that while the Ottoman rulers *did* dictate the terms of diplomacy vis-à-vis the Europeans, they were surprisingly flexible when it came to accommodating cultural/civilizational differences.

Despite the fact that its territorial borders neighboured *both* the Western and Eastern worlds, this neither resulted in a 'clash of civilizations', nor did it prove to be an obstacle to establishing some sort of political order. This holds important implications for the conventional emphasis on civilizational homogeneity as a prerequisite for some form of international order.

By way of concluding, and with reference to the contemporary 'clash of civilizations' thesis between the West and Islam (Huntington 1993, 1996), the weight of empirical evidence discussed in this chapter indicates that the 'inclusion' of an Islamic empire in (European) international society needs to be analysed within a broader historical context, rather than on the basis of the nineteenth-century Eurocentric standards of 'civilization'. The interactions of European and non-European polities before the expansion of European international society shows that there was a process of collaboration and acculturation as well as conflict of their relevant cultural systems. In the twenty-first century, such an historical claim is too important to be overlooked, given the challenges of post-9/11 global politics.

Notes

1 The name Istanbul was not officially and exclusively used until the 1930s. The result of this has been a confusion and interchangeable use of Constantinople and Istanbul (particularly by Western scholars). In this chapter, I use the historical name of Istanbul, Constantinople, for the sake of consistency.

2 The concept of international society can be found in the classic writings of Hedley Bull, Martin Wight, Adam Watson and John Vincent as 'the legitimate founders' of the English School of International Relations (Dunne 1998: 15; Linklater and Suganami 2006: 41–42). These scholars characterized a society of states that originated in Europe and was based on common norms, values and institutions as identified by Bull (Linklater and Suganami 2006: 108–13).

3 This classic differentiation was based on Hedley Bull's classification of system and society in IR literature. According to Bull, '(a) *system of states* (or international system) is formed when two or more states have sufficient contact between them, and have sufficient impact on one another's decisions, to cause them to behave – at least in some measure – as parts of a whole' (Bull 1995: 9). Furthermore, in his analysis, the definition of international society refers to a society of sovereign states. '*A society of states* (or international society) exists when a group of states, conscious of certain common interests and common values, form a society in the sense that they conceive themselves to be bound by a common set of rules in their relations with one another, and share in the working of common institutions'. He then argues that certain common interests, common values and certain rules bind the members of an international society. 'At the same time they co-operate in the working of institutions such as the forms of procedures of international law, the machinery of diplomacy and general international organisation, and the customs conventions of war' (Bull 1995: 13).

4 The official name of the Ottoman state (*Osmanlı devleti*) was *Devlet-i 'Aliyye Osmaniyye* (Sublime Ottoman State – translated as the Sublime Porte into English). It is generally agreed by historians that the people known as the Ottomans (*Osmanlı* or Ottoman Turks) settled in Asia Minor (*Anadolu* – Anatolia) in the mid-1200s as a Turkic tribal group led by Ertuğrul Gazi (1198–1281) (İnalcık 2010; Ortaylı 2007: 13). Amongst his three sons, Osman Bey (Osman I, 1258–1326)

declared his independence from the Seljuk sultanate. Although the foundation of the *Osmanlı Beyliği* is generally accepted as 1299 there is no agreement on the exact establishment date of the *Osmanlı Beyliği* as an independent principality. While Uyar and Erickson (2009: 11) suggest 1301, İnalcık (2010: 17) suggests 27 July 1302 as the correct foundation date of the state. Although the term *Osmanlı* – Ottoman – originated with *Osman Bey* it was his son Orhan Gazi (1281–1360) who transformed a principality into a 'sovereign entity' by capturing the city of Bursa in 1326 and İznik in 1330 in a victorious battle against the Byzantine Empire (please note that the term 'sovereignty' is not used in the sense of European sovereign states here).

3 Curious and exotic encounters

Europeans as supplicants in the Chinese Imperium, 1513–1793

Yongjin Zhang

Introduction

In receiving Lord Macartney's mission in 1793 in the Imperial Summer Palace in Chengde outside Beijing, the ageing Emperor Qianlong, already in his early eighties, wrote a poem, part of which reads:

> Formerly Portugal presented tribute;
> Now England is paying homage.
> They have out-travelled Shu-hai and Heng-chang;
> My ancestor's merit and virtue must have reached their distant shores.
> Though their tribute is commonplace, my heart proves sincerely.
> Curios and the boasted ingenuity of their device I prize not.
> Though what they bring is meagre, yet,
> In my kindness to men from afar I make generous return,
> Wanting to preserve my good health and power.
>
> (Hsu 1990: 159)

Three ideas to which the ageing Chinese emperor alluded here are important for our considerations. First, the Qing Imperium was a Chinese world order presided over by Imperial China and its Son of Heaven, while also being open to the participation of non-Chinese peoples and states. Second, one of the central institutional constructs of this *Pax Sinica* was the tribute system, which regulated and made possible such participation. Third, Europe, and in the first instance Portugal, came to participate in this *Pax Sinica* as one of many tribute-bearing states. What the ageing emperor could not foretell is the symbolic significance of the value that the Macartney Mission represented and its importance in shaping the history of Imperial China's relations with Britain and with an expanding European international society, not to mention the coming clashes between two international orders and civilizations, Chinese and European, in the mid-nineteenth century.

Emperor Qianlong's rejection of the Macartney Mission, in the words of James Hevia (1995: 231), is symbolic of 'the Qing Court's refusal to allow British penetration of China in British terms'. With hindsight, Lord

Macartney's ill-fated mission signals unquestionably the beginning of an end of an historical pattern of interactions between China and Europe. The existing literature and historiography of China's *international* relations in both English and Chinese, have, for different political and pedagogical reasons, concentrated overwhelmingly on what might be called the 'post-Macartney' period, particularly after the Opium War of 1839–42, when the terms of China's international engagement were dictated by the expanding European international society. China's *international* relations, in other words, started with the Opium War, when the West brought the *international* to the Chinese world. Traditional China had *foreign* relations, not *international* relations (Teng and Fairbank 1954; Fairbank 1968). Historical accounts of interactions between China and Europe in the period between the sixteenth and nineteenth centuries have been conveniently consigned to the history of intellectual and cultural exchanges (see, for example, Zhou 2010; Zhang 2009; Mungello 1999; Waley-Cohen 1999; Zhou 1987). They rarely feature in any meaningful way in the International Relations (IR) literature concerning China. This is both an anomaly and a puzzle.

What underlies this problematic anomaly in the existing literature is a subterranean Eurocentrism in the interpretation of world history, as other chapters in this book also contend. This is true of even international theories that put particular emphasis on the historical context of theorizing. Take, for example, the grand historical narratives about the expansion of European international society articulated by the English School scholars (Bull and Watson 1984a; Buzan and Little 2010). In these narratives, the European expansion into East Asia becomes meaningful only when the European society of states globalized the Westphalian system of sovereign states in East Asia and when the European standard of 'civilization' was imposed on China after 1842, and on Japan after 1853 (Gong 1984a; Suzuki 2005). It is curiously silent about the historical relationship between Europe (the West) and East Asia (the rest) before the West established its global domination, when it was the Chinese and other non-European actors that regularly dictated terms of engagement for European states and non-state agents alike in various regional international orders beyond Europe. Existing IR scholarship pays insufficient attention, if any, to a long historical period of sustained interaction between Europe and East Asia before the beginning of the nineteenth century when East Asian states, China and Japan in particular, exercised considerable agency in defining norms and principles of international orders in their own regions (as Suzuki's chapter in this book also shows).

For nearly three centuries after the first group of Portuguese traders ventured into the Chinese waters and landed on the territory of Imperial China in the early sixteenth century, the interactions between China and Europe at the systemic and societal levels – economic, social and even political/military – were sustained and significant. Europeans – be they missionaries (the Jesuits, the Augustinians, the Dominicans and the Franciscans) or traders (the Portuguese, the Spanish, the Dutch and the British) – sought to participate in

Pax Sinica with determination, diligence and enthusiasm. Sometimes this participation took place on terms of parity and equality, but more often Europeans had to accept, albeit reluctantly, inferiority vis-à-vis the Chinese within the social structure of the hierarchy of the Chinese order. The penetration of exotic Chinese goods and the diffusion of the images and ideas of China as a curious land into Europe in the seventeenth and eighteenth centuries both also stand as testimony of such interactions (Zhang 2009). Even after the formation of the Westphalian system in Europe in 1648 and with accelerated European expansion into East Asia, the two international societies, Chinese and European, enjoyed a kind of peaceful co-existence for almost 200 years until the Opium War in 1839–42.

This chapter provides a brief analytical account of the experience of three groups of Europeans – namely, pioneering Jesuits, warriors and merchants, and embassies and diplomats – and their encounters with and participation in the Chinese Imperium before the rise of the West. More specifically, this chapter refers to the period between 1513 (when the first Portuguese reached the Chinese shores and landed near Canton) and 1793 (when Lord Macartney returned to London after his ill-fated China embassy). If the Jesuits were pioneers in cultural and social engagement in Imperial China as a 'curious' land, European traders sought to establish sustained economic and trade exchanges with the exotic empire to exploit its immense wealth. European diplomatic initiatives, on the other hand, accompanied the endeavour of these cultural and economic agents and sought to support them. The examinations of these European attempts highlight a variety of approaches in early European encounters with rich intellectual and cultural traditions of Imperial China, its social customs and institutions, its intricate trading institutions and regulations, and its diplomatic tradition and practices.

I seek to advance three key arguments in this chapter. First, the European expansion into China started in the sixteenth century not the nineteenth century, which was accompanied by notable contestations, mediations and negotiations between Chinese and European worldviews. During this period of European expansion, the two world orders, Chinese and European, seem to have managed a sustained and prolonged period of peaceful coexistence, with only isolated instances of violent conflict between the sixteenth and the nineteenth centuries. Second, there is no historical evidence to suggest that in the period under consideration, the European society of states made any conscious efforts to draw Imperial China into the international order of its own construction. On the contrary, European state and non-state agents participated in the East Asian order primarily defined by local interests, institutions and agendas. Their participation plays a significant role in reinforcing and reproducing the hierarchical social structure of the Chinese world order in this period. Third, such political, cultural and commercial participation by both state and non-state European agents in the Imperial Chinese political, social and economic orders is highly contingent upon the Europeans accommodating, adapting to and/or accepting norms, values, rules and institutions in the prevailing Chinese world order. In other words, it is *Imperial China* that dictates the terms of engagement. In this light, the Macartney Mission in 1792–93, as the first British embassy to China, can be regarded as among the last attempts by the Europeans as supplicants to seek their participation in the institutional order presided over by Imperial China. This is in sharp contrast to the pattern of engagement and conflict between China and the European international society in the post-Macartney period, when the latter imposed by force and coercion the Westphalian institutions on East Asia, dismantling in the process the traditional East Asian international order (Hsu 1960; Zhang 1991; Wang 2005).

Pioneering Jesuits

The contribution of the Jesuits as cultural agents to the intellectual and cultural exchanges between China and Europe in the seventeenth and eighteenth centuries has been widely acknowledged (Rowbotham 1942; Treadgold 1973; Young 1983; Mungello 1989). The lasting impact that the Jesuits have exerted is to be found more as an intellectual venture rather than as a spiritual one. The Jesuits, for example, provided Europe with the first substantive information about Chinese culture and society and made an indispensable contribution to the emergence of Sinology – the study of China – as a discipline in Europe (Millar 2007; Zhang 2009). They also introduced European science and Western learning into China (Peterson 1998; Du and Han 1993). As the first group of Europeans who sought sustained social contact with and penetration into the Chinese Imperium, the difficulties presented to the Jesuits were formidable, as they were confronted with an advanced culture such as they had never seen before. Imperial China's social structure and the Confucian societal configuration made such an accommodation even more demanding. Accommodating Christianity to indigenous culture took on a completely different meaning in this social and historical context.

The secret of the Jesuits' success, if any, lies in their attempts to participate in Imperial China's social, political and religious orders through accommodation, adaptation and even integration before carrying out their Christian mission of conversion. The genesis of the Jesuit policy of accommodation in East Asia is often traced back to St Francis Xavier, one of the founding members of the Jesuit order. Xavier's experience in Japan in 1549–52, in particular, convinced him not only of the need for the Jesuits to learn to speak, read and write in native language and to compromise with local culture, but also the imperative to have great talents among the Jesuit missionaries so that they could meet the locals on equal terms intellectually, if the mission were to be successful. 'It is self-evident', he wrote:

> What we want here are powerful intellects, practised in dialectics, gifted with a popular eloquence, quick to follow error in its shiftings and even to anticipate them, able to snatch the mask from lies which plausibly bear

> the semblance of reality, to unravel sophisticated arguments, and to show the incoherence and mutual contradiction of false doctrine.
>
> (Quoted in Young 1980: 5)

Similarly, Alesandro Valigano, as 'Visitor of all Jesuit missions in the Far East' in the 1570s, advocated a strategy of penetrating China by 'a confrontation with the intellectual aristocracy on its own level of language, social customs and superior talent' (Treadgold 1973: 8).

The key to the formulation of such an approach is the Jesuits' realistic assessment that Imperial China, as the 'Mightie Kingdome' (Lach 1965), was in many ways equal and even superior to Europe both intellectually and materially. The Jesuits, as D.E. Mungello (2005: 81–82) writes:

> [r]ecognized that the Chinese, unlike those in other technologically or materially less advanced parts of the world, could not be converted by overawing them by the European might. Rather, the Chinese needed to be approached as intellectual equals and shown through sophisticated arguments that Christianity was in harmony with some of their more fundamental beliefs.

Matteo Ricci, who arrived in China in 1583 and is said to be China's first immigrant from Europe in the modern sense, was universally regarded as the pioneer in formulating and practising the Jesuits' accommodative approach to Imperial China. Clearly acknowledging the advanced nature of the Chinese civilization, he remarked in his journal that 'of all the pagan sects known to Europe, I know of no people who fell into fewer errors in the early ages of their antiquity than did the Chinese' (quoted in Treadgold 1973: 12). In practice, Ricci's compromise with local culture started with him taking up a Chinese name, dressing in the traditional silk robes commonly used by Chinese literati, being carried about in sedan chairs, hiring domestic servants and even letting his beard grow. In order to approach the literati on terms of equality, Ricci learned to master Confucian classics. He not only spoke Chinese, but also read and wrote classical Chinese, the literary language used by the Chinese literati, the crucial social group that Ricci came to believe held the key to the success of the Jesuits' attempt to convert the Chinese to Christianity. Ricci proudly called himself a *Xi Ru* (西儒, a Confucian scholar from the West). Adopting the lifestyle and identity of Chinese literati was of more than symbolic value. It helped Ricci to develop the Jesuits' accommodative approach in close collaboration with the Chinese literati, the Jesuits' closest counterpart in terms of education, social standing and moral cultivation, who were the most respected in Chinese society and the most influential in the Chinese bureaucracy (Young 1983; Mungello 1989).

Ricci's accommodative approach had several more sophisticated thrusts. As a learned man from the Europe of High Renaissance, Ricci made good use of his knowledge of European science for the benefit of the Jesuits' missionary enterprise in China. Using European cartography, Ricci produced the first map of the world with all place names given in Chinese. To accommodate the Chinese view of their country as the Middle Kingdom, Ricci adeptly placed China near the centre of the map. In close collaboration with Xu Guangqi – one of his high-profile converts, who was credited with having pioneered Imperial Ming's knowledge formation about the West – Ricci produced translated works in Chinese on Western geography, astronomy and mathematics, including in particular the first six books of Euclid's *Elements of Geometry*.

Ricci took another ingenious and compromising approach to choosing the Chinese characters to be used in his translation of the Christian monotheistic God. He was convinced that the two Chinese characters *Shangdi* (上帝 Lord on High) in the traditional Chinese texts could be retained conceptually and literally in his translation referring to God. In his famous essay *Tianzhu shiyi* (天主实义, 'The True Meaning of the Lord of Heaven'), the first significant Jesuit text in Chinese, published in 1603, Ricci was explicit. In his words, 'The God of our country is the same as Chinese *Shangdi*', and 'Our God is the Shangdi in Chinese classics' (Huang 1987: 58). He also came up with a new term, *Tianzhu* (Lord of Heaven), to be used by missionaries and their Chinese converts 'to avoid cultural overlays of *Shangdi*' (Mungello 1989).

Taking advantage of the relative intellectual openness in the unusual social, cultural and political contexts embodied in 'the syncretic spirit of the Ming', Ricci and other Jesuit pioneers embarked on an effort to create a Confucian–Christian synthesis.[1] In *Tianzhu shiyi*, regarded by Ricci (1985) as a 'pre-evangelical dialogue', the main purpose of the narrative was to engage the Chinese literati in philosophical, rather than religious, discussions. Ricci adapted many passages and arguments from the highly esteemed canonical Confucian classics to illustrate how Christianity and Confucianism were compatible and complementary to each other. In the same spirit, Xu Guangqi reformulated the role of Christianity in China as 'supplementing Confucianism and fighting against Buddhism' (补儒易佛, *bu ru yi fo*) (Liu 2008: 477–78).

The use of *Shangdi* would be later entangled in the so-called Chinese Rites Controversy (Minamiki 1985). The central issue of this controversy concerns, however, rituals that Chinese performed in honour of their ancestors as well as Confucius. While Ricci and the Jesuits were prepared to accept that certain rites to ancestors were superstitious, they argued that these rites had important social and moral meanings, which did not violate the monotheistic nature of the Christian God. This proved most controversial among missionaries in China. It led ultimately to Pope Clement XI's Constitution of Prohibitions on the Chinese Rites in 1704, which 'hit the sunshine of royal approbation and finally drowned the Church in a deluge of destruction' (Rowbotham 1942: 116).

Ricci's ultimate achievement is very often seen as obtaining the permission of the Imperial Court of the Ming Dynasty for the establishment of the first European missionary residence ever in the capital of Beijing in 1601. Though

Ricci never secured an audience with the emperor, he was granted freedom of movement in the Forbidden City, the emperor's residence. This was no mean achievement. A more lasting achievement was, however, that Ricci's accommodative approach provided a benchmark against which the Jesuits and other missionaries would be accepted and tolerated (or not) within the Chinese Imperium. Significantly, this benchmark survived the collapse of the Ming Dynasty and persisted when cultural conditions in China became less fertile for Christianity in the period leading to the final showdown of the Chinese Rites Controversy in the early eighteenth century (Minamiki 1985). Emperor Kangxi, whose reign spans a 60-year period between 1662 and 1722, regarded Ricci's practice as setting up a tradition. In remarks made to Messabarba, who led the second Papal Legation to China in 1720, the Emperor said:

> If you want to discuss the doctrine of China, you must necessarily penetrate deeply into the Chinese literary style, and study Chinese literature thoroughly: only then will you be able to argue. I, the Emperor, do not know any Western tongues, and therefore I do not discuss anything Western.
>
> (Huang 1987: 63)

Not surprisingly, Emperor Kangxi issued numerous decrees in which all missionaries in China were urged to follow the practice of Ricci in order to obtain imperial protection (Liu 2002). For Kangxi, 'the practice of Ricci did not imply merely a kind of method, a set of means, to achieve a certain purpose, but a basic attitude of respect for Chinese civilization and culture, obedience to Chinese laws and a following social customs like a native Chinese' (Huang 1987: 58).

It should be properly acknowledged that the Jesuit and other missionaries were accepted and tolerated within the Chinese Imperium also because of the useful service they could render to Imperial China using their secular knowledge. In the extremely violent and chaotic transition from the Ming to the Qing dynasties in the mid-seventeenth century and the unpredictable unfolding of episodes of the Rites Controversy involving changes of court politics, power and policies, the Jesuit missionaries – Adam Schall and Ferdinand Verbiest among them – remained dominant in the Imperial Bureau of Astronomy for 150 years after 1629, as their advanced knowledge of astronomy and mathematics was exploited by successive imperial rulers to make imperial calendars and enhance the legitimacy of their rule. Some astronomical instruments the Jesuit missionaries made, including a quadrant, a sextant and a celestial globe, are still on exhibition in Beijing today. The so-called court Jesuits – those Jesuits who served in the imperial court of the Qing – also provided services as the architects of the Imperial Summer Palace (*Yuan Ming Yuan*, 圆明园), as military engineers, as court painters and as court interpreters for visiting European embassies, and for Imperial China's negotiations with Russia for the Treaty of Nerchinsk in 1689 (Sebes 1961) and the Treaty of Khiakhta in 1727. The Jesuits, Waley-Cohen (1999: 106) writes, 'worked extremely hard and with considerable success to satisfy imperial demands for both aesthetic pleasure and practical science and technology'. This partially explains why several Qing emperors in the seventeenth and eighteenth centuries – Emperor Shunzhi and Emperor Kangxi, in particular – all had close relationships with the so-called 'court Jesuits' (Rowbotham 1942).

It is probably true that in their encounter with Imperial China as a 'curious land', the Jesuits sought ultimately 'to change China into something acceptable to them, to make China partake of Western values' (Spence 1980: 5). In this initial encounter under our considerations here, however, the Jesuits found that they had first to make themselves acceptable in the Chinese Imperium by adopting China's social customs and to accommodate the traditional Confucian values into Christianity. As the outcome of the Chinese Rites Controversy ultimately testifies, the terms of engagement were not dictated by the Vatican but by Beijing. The intervention of the Vatican and the insertion of the international personality of the Papacy in the Chinese Rites Controversy not only led to the abrupt end of the Jesuit encounter, but the banning of Christianity in China by the Yongzheng Emperor in 1724 (Gu 2002).

Warriors and merchants

The Jesuits' intellectual, spiritual and social encounters with the Chinese Imperium in the seventeenth and the eighteenth centuries had been preceded and were accompanied by activities engaged by two other groups of Europeans, warriors and merchants in search of exotic spices and wealth. As Geoffrey Hudson (1965: 236) writes, 'when the Portuguese first reached China from Malacca [in 1514], there was no thought except for material commerce and for an arrangement with the Ming emperor to secure this'. Like the Jesuits, European warriors and traders found the Chinese 'a people of great skill' and 'on a par with ourselves' in their very early encounters (Bitterli 1989: 134). They were struck by 'the richness of its commerce, the infinite number of its revenues'. European rulers would, they believed, 'feel very small before this great Monarch' (Spence 1980: 43). They were also 'favourably impressed with the orderly management of trade' (Wills 1984: 37). Unlike the Jesuits, however, their first-hand experience was not primarily with the cultural and intellectual stalwarts of traditional China, but with the greatest single economy in the world in the sixteenth and the seventeenth centuries; an elaborate regional trading network in maritime Asia; a strong military power; and a highly bureaucratized state which tightly controlled and regulated Imperial China's trade. Though their knowledge about China was neither sufficient nor accurate at the time, when the first Portuguese ships sailed into the mouth of the Pearl River in 1513–17, Europeans were under no illusion that Europe was in any way superior to China in terms of commerce, wealth or military power.

As the earliest European traders to reach the Chinese shores, the Portuguese had sought to participate in an elaborate existing regional trading system in maritime Asia – China, Japan, Taiwan, South-East Asia and India, among others – rather than attempting to reinvent one. The Portuguese capture of Malacca in 1511 was a significant step in this cross-cultural encounter. Wills (1998: 334) describes the beginning of the Portuguese participation:

> In [the] Cheng-te [Zhengde] period [of the Ming, 1505–21], ships from Southeast Asian tributary states were allowed to come as frequently as they wished, without regard for the limitations of time and number specified in the regulations of the tribute system and their trade was taxed ... This Southeast Asian trade, officially approved, but in violation of basic rules of the tribute system, provided the matrix for the flourishing trade between Siam and Melaka and South China, within which matrix the Portuguese began their relations with China.

The early attempts by the Portuguese to participate in Chinese trade were full of drama, and not free from violent military conflicts (Zhang 2005). Appearing at the mouth of the Pearl River near Canton unannounced in 1513, the Portuguese were allowed to trade on board their ships, but not to proceed to Canton (Wan 2001; Zhang 2005). The first Portuguese embassy, regarded as an unknown tribute bearer from the Great West by the Ming court in Beijing, was subsequently permitted to proceed from Canton to Beijing in 1520. Yet, from the very beginning, the Portuguese attempts were doomed to fail because of mutual misperceptions as well as the Ming court's displeasure about the Portuguese capture of Malacca, a loyal tributary state to the Ming, and the reckless behaviour of the Portuguese sailors and soldiers. Simao Peres d'Andrade, who commanded the Portuguese ships in their second expeditions to Canton in 1519, was noted to have 'refused to pay the Chinese the usual customs duties, administered justice according his whims, and began building fortifications in the teeth of opposition' (Bitterli 1989: 136). The stories of the Portuguese abduction and kidnapping of young children, which turned out to be partially true, did irreparable damage to early Portuguese efforts. The Chinese engaged the Portuguese in two sea battles in 1521–22, in Tunmen and Xicaowan near Canton, driving away the Portuguese in both cases (Wan 2001; Wills 1998). After 1522, the Portuguese were officially banned from trading with China for more than 30 years until 1554.

The 'Sino–Portuguese War' of 1521–22 is regarded by some Chinese historians as the 'first military conflict between China and the West' (Wang and Pan 2004). Significantly, few serious direct military clashes took place between Imperial China and maritime European powers over the next three centuries until the Anglo–Chinese Opium War in 1839, while the Portuguese, the Dutch, the Spanish and the British were all engaged in, and occasionally fought each other for the flourishing China trade. Europeans, however, also took part in the imperial Chinese order in a different capacity, as warriors of a different nature. Three cases are particularly noteworthy.

The first is related to the court Jesuits. Two examples would suffice here. In 1622, when Macao was under attack by the Dutch, Father Adam Schall provided tactical and technological assistance to its defenders. In 1642, at the request of the Ming emperor, Schall helped cast European-style cannon and provided information on European military technology. Schall's advice was also sought by the emperor for the defence of Beijing against the onslaught by the rebel Li Zicheng's forces. Father Ferdinand Verbiest, who succeeded Father Adam Schall as the Director of the Imperial Bureau of Astronomy, was also skilled in the art of casting cannon. From 1660 to 1688, while serving the Qing court as an imperial astronomer, Verbiest was also responsible for the casting of 566 cannon (Shu 1994; Witek 1996).

The second case involved the ironic return of Portuguese warriors in the twilight years of the Ming Dynasty, over 100 years after their defeat in the sea battles near Canton. The difference is that this time the Portuguese soldiers were involved in defending the Ming. Between 1621 and 1647, the Portuguese embarked on a number of military expeditions at the request of the Ming court to assist the Ming in their military contest with the Manchu forces (Dong and Huang 2009). A small group of Portuguese artillery men were brought from Macao to Beijing to train the Ming soldiers how to use cannon (Wills 1998: 352). Even as late as 1645, one year after the death of the last Ming emperor, the Jesuit Father Francisco Sambiasi arrived at Macao from Nanjing with further requests for Portuguese military aid in exchange for new Ming concessions. The Portuguese initiatives to send military aid and auxiliaries to assist the Ming forces were clearly 'an attempt to stabilise any further deterioration in Portuguese relations with the Ming' (Souza 1986: 198). Wittingly or unwittingly, the Portuguese were embroiled in the internal politics in the dynastic transition between the Ming and the Qing.

The third case concerns the Dutch military cooperation with the Qing government, which needs a more detailed elaboration. Though the Dutch East India Company (VOC) was only formed in 1601, '[t]he Dutch became enthusiastic participants in the "China trade" almost as soon as they arrived in Asian waters in the early seventeenth century' (Atwell 1998: 396). Like the Portuguese and other Europeans, the Dutch also believed that 'they could make more profits by inserting themselves into already existing systems of trade and taxation' (Keene 2002: 76). A series of high-handed approaches and tactics, sometimes resorting to the use of force, were adopted by the VOC in the early decades of the seventeenth century to establish a foothold on the Chinese coast for the company's China trade. The governor-general of the VOC, Jan Pieterszoon Coen, once remarked that company ships should 'pester and hassle the whole coast of China to the maximum degree possible, so that the Chinese will be forced to come to a negotiated settlement [with us], which will undoubtedly happen' (Kops 2002: 539).

The main thrust of Dutch diplomacy involved offering Dutch military cooperation and naval assistance. This strategy was principally motivated by the VOC's long-cherished desire to secure possible trade concessions from the

Qing government, such as a long-term trade agreement, in carrying out its China trade. It must also be understood against a complex background of what Fernand Braudel (1994) calls 'a global struggle between the Dutch and the Portuguese' and the established practices of Dutch diplomacy in East Asia and its recent successes. The Dutch naval assistance to Raja Sinha II against the Portuguese helped the conclusion of the Kandyan Treaty of 1638 (Keene 2002: 79–80). Also in 1638, the VOC military assistance to the *shōgun* in his quest to pacify Japanese Catholics won the company some trade concessions, though carefully restricted (Kops 2002: 539).[2] At the same time, the newly established Manchu government, in need of consolidating its conquest of the Ming, became open to considering the Dutch overture for military cooperation and assistance to defeat the remnants of the Ming loyalists. Among them was Zheng Chenggong (Coxinga), whose forces took over Taiwan from the Dutch in 1662, and whose large and well-organized fleet of both cargo and war junks had posed a serious threat to the Dutch position and interests in China–Japan trade. There was therefore a convergence of interests between the Qing government and the VOC in defeating Zheng.

As early as 1655, two ambassadors of the first Dutch Embassy to Beijing sent by the VOC had received secret instructions to explore the possibility of a naval alliance between the VOC and the Qing government against Zheng and his fleet. The VOC could assist the Qing with 'ships and personnel' and 'could put so much pressure on [Coxinga], that he would fall into the hands of the Tartars'. However, the two ambassadors were also instructed to ask what trade privileges the Qing government 'would be willing to grant us' (Kops 2002: 546). It was only in 1662, after the fall of Taiwan, however, that the VOC sent 12 ships to the Chinese coast to attack Zheng's outposts and shipping in order to 'restore the company's reputation'. The Dutch also opened negotiations for military cooperation with Qing officials in Fujian province. An imperial edict subsequently arrived in Fuzhou authorizing a joint naval campaign with the Dutch against Zheng's forces, and granting the Dutch the privilege to trade every year and to build a trading station in Fuzhou. However, the edict was delivered by an imperial envoy from Beijing only after the Dutch fleet had already left. In the ensuing 20 years, the Dutch entered into a number of negotiations with the Qing government both in Beijing and at various coastal locations on trading privileges in exchange for Dutch military assistance. To the frustration of the Dutch, the Chinese offers, if any, seemed to be always less than satisfactory, and constantly subject to different interpretations (Wills 1968: 231–41; Wills 1998).

The sustained endeavour of the VOC with the backing of its considerable military power had an ambitious goal to secure a binding agreement with the Qing government for the Dutch trading privileges in China. It failed, not because of the inherent lack of merits in this approach, but rather because European traders were confronted by and had to work with highly regulated and heavily guarded trade regimes when they inserted themselves into an established trading network in East Asia.[3] European traders, Rahman (2003) writes, 'followed the same trade routes, used the same ports and exchanged the same genre of products. They first began to use the China trade to procure products for Europe, but what proved more lasting and lucrative was the regional trade (or the country trade) involving Asian markets and trades'. In so doing, European traders entered effectively a 'controlled relationship', where they:

> found their freedom of action limited both geographically and socially. It was the Chinese and the Japanese who decided where the two cultures meet, who should take part and which outside influence would be admitted. Subject to drastic controls, trade was tolerated, even welcomed; but the hosts made it plain that they did not need foreign trade and wanted to decide the terms of the encounter.
>
> (Bitterli 1989: 133–34)

Nowhere is such a 'controlled relationship' more compellingly demonstrated than in the construction of Macao as a Portuguese enclave in the late sixteenth and early seventeenth centuries, and in the implementation of the Canton system in the eighteenth century in managing European presence in Imperial China. In the words of Hudson (1965: 238):

> The new race of sea barbarians who had appeared on the horizon of the empire could not be admitted to imperial favour after they had shown such truculence or disposition, nor could they be allowed to penetrate into the interior of China, but as long as there was profit to be obtained from dealings with them, they might be tolerated in certain ports.

Whether or not 'Macao was given to Portugal for the services rendered by Portuguese ships to the long suffering Chinese navy in dealing with the pirates in the mid-1550' (Hudson 1965: 238) remains contentious even today (Wan 2001: 94–103; Zhang 2004). It is nevertheless clear that allowing the Portuguese to build permanent settlement on the peninsula of Macao in 1557 was a local solution to accommodate increasing numbers of Portuguese traders along the Canton coast. When the official ban of the Portuguese traders was relaxed at the beginning of the 1550s, the purpose of this local solution was to confine the Portuguese traders in an isolated place on the periphery of the Chinese empire. There was clearly no territorial concession on the part of China. Not only did the Portuguese continue to pay ground rent, but they also had to pay customs duties and harbour dues. Macao remained completely integrated into the Ming administrative systems, with a Ming garrison (Huang 2003; Tang 1999).

The Chinese also exercised their authority to forbid the Portuguese from fortifying Macao. Even after the failed Dutch military attack on Macao in 1622, only limited fortification of Macao was allowed by the Chinese, who could also revoke such concession at will. Drastic restrictions of freedom of movement were placed on the Portuguese settlers, when in 1573 a wall and a gate, the *Porta do Cerco* (Circle Gate), manned by guards, were erected and the Portuguese and other foreigners were forbidden to go beyond it without permission. This *Porta do Cerco*:

> … was opened chiefly when markets were held in Macao, about once a week; afterwards the door was again locked and sealed with strips of paper bearing official stamps. Since this was the only access to the mainland, and the channel through which the town was supplied with food, the Chinese could put the Portuguese under pressure, for whatever reason, simply by ordering the *Porta do Cerco* be shut.
>
> (Bitterli 1989: 143)

What Wills (1998: 351) calls 'Macao's charter of survival through submission' took on a new shape in 1614, when the governor-general of Guangdong and Guangxi sent officials to proclaim a full set of regulations. These consisted of the following five points: 1 Macao must not harbour Japanese; 2 The buying of Chinese people is forbidden; 3 All ships, including warships, must pay duties and must come into Macao's Inner Harbour. Anchoring and trading in the outer islands is strictly forbidden; 4 Trade must be conducted in Canton, not at Macao, and duties on goods must be paid there; 5 New construction on Macao is strictly forbidden, but old structures may be repaired or rebuilt to match their previous condition. The Portuguese were urged to obey these regulations to the letter. The five points were engraved on a stone tablet that was set up in front of the hall of the Loyal Senate in 1617 (see also Tang 1999). 'These regulations, and their revisions and expansions in the 1740s, were fundamental to Chinese policy toward Macao down to the nineteenth century' (Wills 2011: 48).

The Portuguese acquiesced with these onerous regulations for two principal reasons. First, even during years when such drastic restrictions were instituted, the China trade carried out by the Portuguese prospered. In one estimate, '[t]he half century from 1570 to 1620 was generally one of peace and prosperity on the south China coast. Portuguese ships carried Chinese silks and other goods to Japan and returned with Japanese silver, perhaps 400,000 taels per year in the 1580s, 1,000,000 or more in the early 1600s' (Wills 1979: 213). In other words, profit and prosperity of trade, not 'freedom of trade' or 'freedom of movement' was the goal of the Portuguese in Macao. Second, the municipal government of Macao – a self-governing body established in 1595 and dominated by a resident Portuguese merchant oligarchy – clearly knew that as the Portuguese enclave off the Chinese coast, Macao was completely at the mercy of the Chinese state. '[T]he indignant trumpeting of captains-major and captains-general about Portuguese honor and craven submission to the mandarins' (Wills 1998: 347) would not change this fact.

As in the case of constructing Macao as a Portuguese enclave, the Imperial Chinese state established the so-called Canton system in the eighteenth century as another attempt to control European – particularly British – trade with Imperial China and to limit foreign contact to border areas. The evolution of the Canton system started with an imperial decree in 1759 explicitly ordering that Canton be made the only port opened to foreign commerce. Five Regulations for Dealing with Foreigners (防夷五事), first issued in 1760, were expanded and revised in subsequent years into the New Eight Regulations for Dealing with Foreigners (防夷新规八条) in 1835. These regulations constituted the core practices and rules to which European trade in Canton was subject. The so-called New Eight Regulations were, to put it crudely: 1 No foreign warships may sail inside the Bogue [the harbour approach to Canton city]; 2 Neither foreign women nor firearms may be brought into the factories [warehouse complex reserved for foreign traders within the harbour, but outside the walls of Canton city]; 3 Foreign ships should not enter into direct communication with the Chinese people and merchants without the immediate supervision (of a native Chinese); 4 Each factory [each trading nation had its own 'factory'] is restricted to a maximum eight Chinese at its service (irrespective of the number of its occupants); 5 Foreigners living in the factories must not move in and out too frequently, although they may walk freely within 100 yards of their factories; 6 Foreigners may not communicate with Chinese officials except through the proper channel of the Co-hong;[4] 7 Foreign trade must be conducted through the hong merchants; 8 Foreign traders may not say in Canton after the trading season [between October and March] and should return home or go to Macao (Hsu 1999: 201). Needless to say, foreigners living in these 'factories' were subject to Chinese law.

Was the initiation of the Canton system motivated by 'Kangxi emperor's appreciation of the fiscal value of trade and his desire to control it in a centralized, uniform and fiscally rational manner' (Wills 1968: 245–46)? How much did the international trade of China flourish in the eighteenth century under the Canton system? These are intriguing historical questions. They are nevertheless not the central concerns here. It is worth noting, however, first that the Canton system can be regarded as a continuation of Imperial China's efforts to quarantine foreign traders in its border areas; and second that European submission to the Canton system was tenuous and contingent on the delivery of profitability to foreign traders. European complaints against the Canton system in the nineteenth century proved to be one of the contributing factors that led to the Anglo–Chinese Opium War in 1839.

Embassies and diplomats

European trade with China in the fifteenth and sixteenth centuries was carried out, in the words of Geoffrey Hudson (1965: 235), 'unofficially and not by treaty'[5] and 'subject to harassing restrictions'. Hudson also remarked that 'China remained outside the world of diplomatic intercourse formed by the European states and (including also its outer ring) such Asiatic powers as the Ottoman Empire'. While it is true that maritime Europe and Imperial China did not establish formal diplomatic relations until the mid-nineteenth century, this does not mean that Europeans did not actively seek to form such a relationship with Imperial China before the nineteenth century. The Macartney Mission in 1792–93 was preceded by a series of European embassies to Beijing – the Portuguese, the Dutch and the Russian.[6] When Lord Macartney embarked on his intrepid journey of 2,340 miles from Canton to Beijing through rivers and canals, he followed, literally, the well-trodden paths of Dutch embassies and the Portuguese envoys more than a century ago.

In contrast to the intellectual and social encounters of the Jesuits, European diplomatic ventures into Imperial China were not as determined and persistent. Unlike the single-minded and sustained economic engagement of European traders, diplomatic contacts were not only limited, but also tentative and intermittent. The European embassies and diplomatic missions associated with such ventures, as Wills (1984: 4) argued, 'had diverse and distinctly unsystematic origins in the shifts of Ch'ing court politics and coastal policies and in the experience and goals of the Dutch and the Portuguese'. What these European embassies and diplomats encountered, however, was the same well-entrenched tribute system with an intricate ceremonial and institutional complex, the origins and pedigree of which could be traced right back to 600 BC or earlier (Zhang and Buzan 2012). They all had to deal with a time-honoured diplomatic tradition as part of Imperial China's statecraft, which was believed to have successfully managed China's foreign relations for millennia.

The first Portuguese embassy to Imperial China (by the same token, the first European embassy) to open up lucrative trade with China arrived in Canton in 1517. However, it was not until January 1520 that Tome Pires, the designated ambassador of King Manuel of Portugal to the 'king of China', obtained permission to travel to Beijing. Tome Pires and his party did reach Beijing, but he 'was not admitted to an audience with the emperor: the authorities burnt the letter from the King of Portugal, refused to accept presents, and showered the delegation with reproaches' (Bitterli 1989: 136). The first Portuguese embassy did not fail because it mounted any challenge to the ceremonial and ritual order of Imperial China, which formed the core of the Chinese tributary system or to the Chinese diplomatic tradition. As is noted in the Portuguese records, while waiting in Beijing for an audience with the emperor, embassy members 'had to go on the first and fifteenth of every lunar month to prostrate themselves before the wall of the Forbidden City' (Wills 1998: 338). It failed, as discussed earlier, because of a combination of reckless and bellicose Portuguese commanders and soldiers in Canton, the Ming court's displeasure about the Portuguese conquest of Malacca and the Ming court politics prior to the death of Emperor Zhengde.

In addition to the first Portuguese embassy to the Ming in 1520, a cluster of European diplomatic overtures to Imperial China took place in the second half of the seventeenth century, which included three Dutch embassies and two Portuguese missions to Beijing. These are in chronological order, the first Dutch embassy led by Pieter de Goyer and Jacob de Keyser in 1655–57, the second

Dutch embassy headed by Pieter van Hoorn in 1666–68, the Portuguese embassy led by Manoel de Saldanha, 1667–70, the Portuguese mission led by Bento Pereira de Faria, 1678, and finally, the third Dutch embassy headed by Vincent Paats, 1685–87 (Wills 1984; Kops 2002). A Russian embassy also came to Beijing from the north after the conclusion of the Treaty of Nerchinsk in 1689 (Sebes 1961).

Unlike Lord Macartney in Chengde in 1793, none of the European embassies in the seventeenth century in Beijing challenged the ceremonial supremacy of the Son of Heaven, the Chinese emperor, a core assumption and an institutional cornerstone of the Chinese world order. Kowtowing to the emperor seemed to be a non-issue. European embassies, like other non-European embassies from inner Asia and South-East Asia, kowtowed not only to the emperor at the imperial audience. As part of the embassy routine, the Dutch ambassadors and their entourage also kowtowed at the Board of Rites when presenting gifts and the official letter to an 'altar like table' symbolically representing the emperor. The Van Hoorn embassy is noted to have gone to the Board of Rites to practise the ceremonies they would perform one day before the scheduled audience with the emperor. More full kowtows were performed after the audience at the banquets given in the name of the emperor to entertain the embassies and their parties at the Board of Rites (Wills 1984: 25–34).

European envoys seemed to have also conformed, without question, to other ceremonial forms associated with foreign embassies under the Chinese tributary system, carrying the banner of a tributary state in their lead boat and designating their gifts to the emperor as tribute, thus acquiescing in, at least ceremonially, the tributary status of the countries they represented vis-à-vis Imperial China. In return, they enjoyed cordial and relaxed receptions in Beijing.[7] In addition to the formal audience with the emperor, 'Saldanha had two interviews at the Ch'ien-ch'ing Gate [with the emperor]; Pereira de Faria was given a banquet in the imperial presence … , and the "tribute memorial" he brought was read in Chinese, Manchu, and even in Portuguese, Paats had at least one interview with the Emperor, who heard a translation of some kind of memorial he had presented, asked questions, and listened to the European music' (Wills 1984: 32).

How can we explain why European embassies were seemingly unconcerned about complying with the Chinese requests to conform to all these deferential, and even humiliating, ceremonial rituals? One Dutch embassy, as we know from the historical record, had been explicitly ordered to comply with all Chinese requests in order to obtain trade concessions. Is it possible that the Dutch and the Portuguese envoys, unlike their counterparts from inner Asia and maritime Asia, chose not to accept the institutional and symbolic values associated with kowtowing and other ceremonies while performing them? Or is it simply because, at least in the case of the Dutch embassies, merchants as diplomats (van Goor 2004) played a key role and as they 'were servants of a trading company and citizens of a republic, they did not have to worry about the personal honor of a sovereign' (Wills 1968: 249)? Jonathan Spence (1999: 42) was probably right in observing:

> These various embassies and legations, though from proudly independent sovereign states in the West, all adhered to the cumbersome Chinese rituals of deference to the Chinese emperor – including the nine prostrations of the kowtow, and the use of self-demeaning language – that the Chinese considered an essential part of the tributary system of foreign relations. By ignoring their own recently required codes of correct diplomatic contact between sovereign nations, in order to achieve their short-term goals, the Westerners were unwittingly shoring up the Qing Court's views of Chinese superiority.

By submitting themselves to the ritual order of Imperial China, these European embassies, therefore, helped produce *Pax Sinica* in East Asia. The maintenance of the pretence of the ceremonial supremacy of the Chinese emperor in all these European diplomatic overtures certainly lent itself to making it possible for European embassies to carry on their missions. However, if their understanding of the centrality of the ritual order in the Chinese Imperium was sound, their knowledge of the role foreign embassies played in Imperial China's tributary system was grossly inadequate. For embassies from inner Asia (including Russia) and East and South-East Asia, trade privileges and concessions associated with the regular tributary missions were what they came to Beijing to request. The European embassies, however, saw it as their principal goal to achieve negotiations in the capital, most desirably directly with the emperor, for trade privileges in and access to Imperial China. What the VOC had hoped for when it dispatched its first embassy to China in 1655 was to have 'an imperial decree – written, sealed, and delivered – that would allow the company unlimited trade with mainland China' (Kops 2002: 570). No more and no less.

Such a European approach was invariably frustrated in the seventeenth and eighteenth centuries because of its incompatibility with Imperial China's longstanding statecraft of managing foreign tributary missions and associated activities as essentially part of a broad security strategy to limit foreign contact, particularly in the capital, and to confine this contact, including trade relations, to the border areas and trading ports, as far from the capital as possible. We have already discussed the effectiveness of such bureaucratic control in the cases of the construction of Macao as a Portuguese enclave, and the operation of the Canton system. Also central to such a management system of bureaucratic control was the strict unilateral bureaucratic regulation of the size and frequency of tributary missions for each tributary state, down to the amount of trade and the number of foreigners coming to the capital. European embassies, like all other non-Chinese embassies, were subject to the same unilateral bureaucratic regulation in this regard.[8]

The biggest pitfall of the seventeenth-century European embassies is perhaps not just that they did not realize 'that embassies were primarily ceremonial', but that 'the court did not ordinarily make substantive decisions about foreign affairs except on the recommendation of the provincial officials involved' (Wills 1998: 336). Nothing better illustrates the ultimate importance of bureaucratic decisions in determining the fate of European embassies by the following imperial edict. While the first Dutch embassy led by Pieter de Goyer and Jacob de Keyser was waiting in Beijing, the emperor sent an edict to the Board of Rites, part of which reads:

> So I order you [president of the board] and the other members, that you people come to a useful decision regarding their [the Dutch] petition which they did through this embassy, to be allowed to go and come in my nation, and that you report back to me on it [the decision].
>
> (Kops 2002: 561)

Such bureaucratic control in terms of policy decision added more woes to the seventeenth-century European diplomatic missions to Imperial China.

While it may be true that 'it was Dutch acceptance of Chinese form of inequality that was primarily responsible for the avoidance of more explicit conflict' (Wills 1968: 249), it is the unilateral bureaucratic control on the part of Imperial China that prevailed over the Dutch assumption of reciprocity in regulating Sino-Dutch relations in the wake of three Dutch embassies. The occasional Dutch references to 'the law of all nations' and 'the custom of all princes' in their negotiations fell on the deaf ears of Imperial Chinese officials. In his discussion of the Dutch embassies, Wills (1968: 248–49) made two insightful observations on muted conflict in the Sino–Dutch diplomatic encounters. First, in the Sino–Dutch negotiations, 'these values (the concept of a community of equal states adhering to a common code for intercourse, a code increasingly formalised in international law) did not come into explicit conflict but remained implicit in assumptions about the rational means to achieve limited practical ends, such as trade and military cooperation'. Second, 'there is no indication that Chinese acceptance of international law and equality among nations was a conscious goal of Dutch policy'. Given the Dutch hegemony in Europe in the seventeenth century, it is significant that in this early encounter between the Chinese and the European international societies, the latter made no explicit attempt to impose European values, customs and law on the former. European interactions with Imperial China at the state level operated according to distinctive Chinese norms and institutions.

Conclusion

The early encounters between East [Asia] and West [Europe], Donald Lach (1965: xii) argues, had a particular characteristic that is often obscured. 'While Europeans dispatched trading, diplomatic, and religious missions to Asia, Asian countries never sent similar missions to Europe on their own initiative'. Further, not only were the relations between East and West 'ordinarily conducted within a framework and on terms established by Asian nations', but also most Europeans were in Asia 'on sufferance'. For almost 300 years between 1513 and 1793, Europeans as either non-state or state agents – cultural, economic and diplomatic – had extensive and sustained, if in some cases also erratic, engagement with the Chinese Imperium, in spite of the formidable and seemingly impregnable barriers erected against foreigners by Ming China, and reconstructed and reinforced by the Qing. In exploring this 'curious and exotic land' for cultural enrichment and economic wealth, Europeans became increasingly significant but invariably marginal participants in the longstanding cultural, economic and imperial order principally, though not exclusively, embodied in the Chinese tributary system. From the sixteenth to the end of the eighteenth centuries, European traders – first the Portuguese, and then the Spanish, the Dutch and the British – were one of many participating groups – smugglers and pirates among them – in the 'China trade'. The economic impact they exerted on either Ming or Qing China – arguably the greatest economy of the world at the time – was, however, insignificant, if not entirely negligible.

In the discussions above, I have used a few examples to highlight a rather obvious, but often obscured point, i.e. that it was Imperial China that unilaterally dictated the terms of engagement. It is also patently clear that the diplomatic dimension of these 'curious and exotic encounters' was the beginning of cross-cultural exchanges between two international societies – Chinese and European – based on completely different concepts and conceptualizations of justice, order, legitimacy, and organizing principles of state and society. Yet, Europeans did not challenge, if ever they did question, the assumptions, worldviews, the legitimacy and predominance of the Chinese world order they encountered. Rather, like other non-Chinese participants in *Pax Sinica*, they accepted, acquiesced in or adjusted themselves to the embedded norms, rules and institutions of the Chinese world order in conducting what we refer to today as 'international relations' with Imperial China.

In other words, in the first 300 years of European expansion into East Asia, early modern Europe, even after its transformation into a society of states post-1648, did not seek to incorporate structurally Imperial China into the European international system/society, nor were Europeans particularly concerned about the normative divergence of the two international orders, nor the terms of engagement and institutional framework largely set out by Imperial China for the meeting of two international societies. This regional order, characterized by Chinese dominance rather than Western domination, thus prevailed for over 300 years after the Portuguese first appeared along the Chinese coast in the early sixteenth century. Two civilizations and two international societies seem to have managed to have a sustained and prolonged period of peaceful coexistence – with only isolated instances of violent conflict. This order came to an end only when the European society of states began to impose its own type of international society on East Asia, and when the European standard of 'civilization' was introduced to enforce a certain kind of cultural unity as membership criterion for an expanding European international society.

In his embassy to Emperor Qianlong's China, Lord Macartney made an insightful comment when he wrote 'Nothing could be more fallacious than to judge of China by any European standard' (Macartney 1962: v). By the time the second British embassy, led by Lord Amherst, was sent to China in 1816, the British had vanquished Napoleon and become Europe's most powerful nation. A profound change was then underway, with the British contemplating the use of European power to impose their will and judgement on Imperial China by the European standard of civilization. On his way back to England, Lord Amherst had a private audience with the now-exiled Napoleon on the island of Saint Helena. On hearing the suggestion that Britain might use force against China, Napoleon remarked, 'You say that you might awe them by means of a maritime argument, and thus force the mandarins to submit to the European etiquette. This idea is madness' (Peyrefitte 1993: 517). As we all know now, it is exactly the practices of what Macartney called 'fallacies' and what Napoleon remarked of as 'the idea of madness' that led to the Opium War in 1839 and came to dominate Europe's relations with China in the century thereafter.

Revisiting this historical experience of cross-cultural exchange and civilizational encounters between China and Europe raises, therefore, three intriguing questions that are of theoretical and practical interest in understanding global international society today. First is the inadequacy and incomplete nature of the grand historical narrative of the expansion of international society articulated by the English School. The analytical historical account provided above is meant to extend the story of expansion of international society, but told from a non-European perspective with temporal emphasis on the period before the age of high imperialism in the nineteenth century. More needs to be done along these lines to correct the Eurocentric nature of the existing grand narrative. Second is the question of how and in what aspects this formative encounter between Chinese and European civilizations and international societies prior to the nineteenth century exercised any influence on the formation of the European international society and its subsequent imperialist expansion into Asia. As Geoffrey Hudson (1965: 236) long ago observed, in the eighteenth century it was China that 'was a great power culturally in Paris than was Europe in Peking', which 'reached out and cast a spell over its future conqueror, leaving indelible traces in the cultural tradition of Europe'. Counter-intuitively, inter-civilizational dialogue of this nature should have had notable influence on the evolution of European international society. Little research has been done, however, on this subject by IR scholars. Carrying out such research would also complement attempts 'to explore the role that the colonies played in the development of the European international society', suggested by Richard Little (Chapter 8) in this volume. Finally, and most fundamentally, it asks the question: how much do cultural differences matter in constructing a lasting, stable and peaceful international order? The historical record examined in this chapter of a prolonged period of relatively peaceful co-existence between the Chinese and the European international societies in a non-European order should give us pause for thought on the claim that *only* cultural unity and civilizational homogeneity could provide a solid foundation for contemporary international society. Instead, embracing cultural and political pluralism of an increasingly globalized world, a critical interrogation of 'cross-civilizational and cross-regional interdependencies' in global international society today (Kayaoglu 2010: 195) is in order.

Notes

1 As Mungello noted, partly to justify their accommodative approach to China in Europe, the Jesuits claimed that Confucianism contained truths derived from the natural world and human reason and lacked only truths of revelation. They also argued that most social and moral truths of Confucianism, such as honouring parents and treating others as we ourselves would wish to be treated, were similar to Christianity. Furthermore, Confucianism, though it lacked divine revelation, was complementary to Christianity and could be used to enrich the teachings of Christianity. See chapter 2 of Mungello 1989; see also Millar 2007.

2 Kops (2002: 546) further noted that, 'Military assistance in exchange for trading privileges also conformed to the VOC's earlier dealings in Asia, not just with the Japanese government, but also with the King of Siam as recently as 1650, so neither the idea of nor the methods to negotiate an alliance were novel'.

3 It is particularly important to note that this was not a problem for European traders only. It was a problem for non-European demand for trade, too. As Wills (1984: 20) notes, 'In the mid-sixteenth century, the Ming mounted no effective diplomatic or military response to Altan's encroachments, and only after thirty years of debate, military expense, and Mongol's raids right down to the wall of Peking, did they work out an arrangement that met the Mongol's rather modest demands for increased trade with China without jeopardizing China's defenses'.

4 Co-hong refers to the guild of Chinese merchants authorized by the Qing government to trade with Western merchants in Canton during the eighteenth and the early nineteenth centuries. It is integral to the Canton system.

5 The exception is perhaps Russia, which did sign the Treaty of Nerchinsk with China in 1689 to regulate its trade relations both at the border areas and in the capital Beijing.

6 In those years, the Vatican also sent a number of envoys to both the Ming and the Qing courts.

7 This is in sharp contrast with the fate of the Portuguese mission to Japan in the mid-seventeenth century. 'The Portuguese were expelled from Japan in 1639 (in the wake of Shimabara Rebellion), and when in 1640 Macao sent an embassy to please for reconsideration, the entire party of officers, merchants and seamen was executed' (Wills 1998: 353).

8 For a discussion of how non-European tributary missions were subject to such bureaucratic control and regulations, see Bielenstein (2005).

4 Europe at the periphery of the Japanese world order

Shogo Suzuki

Introduction

In the English School's interpretation, the history of *societal* relations with the non-European world and the (European) international society typically begins in the late-nineteenth century on the back of European imperialism. The story of Japan's entry into European international society is no exception to this. Prior to 1853, when US Navy Commodore Perry forced Japan to open its borders to the Western powers, Japan is said to have entered a period of 'seclusion' during the reign of the Tokugawa shogunate (1603–1867), refusing all contact with the West, except for Holland.

This narrative, however, is problematic in that it completely casts aside the history of considerable contact between the Europeans and the Japanese. The Japanese had gradually established trading relations with the Europeans since 1543, when the Portuguese arrived at Tanegashima Island. The Spanish arrived in 1584, as did the Dutch in 1600. The first Christian missionary, led by Francisco Xavier, arrived in 1549, and successive missions were to follow. Trading relations quickly followed: the Portuguese had begun trading activities in Nagasaki in 1570, followed by the Dutch in 1609 and the English in 1613. During this time all European powers were required to subject themselves to diplomatic rules as dictated by the Japanese, and this would continue right up to the day that Perry arrived in Japan.

Why has this history of pre-1853 European–Japanese contact been relegated to the background? The answer lies in the entrenched Eurocentrism amongst the English School and other international relations theorists. Concerned with exploring and 'proving' the possibilities of a 'moral' life in an anarchical international order, scholars of the English School tradition have attempted to demonstrate how norms of tolerance and coexistence have (uniquely) evolved in European international history since the Treaty of Westphalia in 1648, which eventually spread throughout the globe (Keene 2002; Kayaoglu 2010). This identification necessarily entailed the formation of a non-European 'Other' that was a passive recipient of European norms. Stories of non-European strength are hardly conducive to the construction of a morally superior European/Western 'Self' willing to spread progressive norms, and as such have been frequently forgotten.

Critiques of such Eurocentric historical narratives have long been articulated in the field of Asian history (Tashiro 1982; Toby 1984; cf. Cohen 1984), but the same cannot quite be said of the English School. While there is now a growing awareness of the approach's blindness to the darker sides of the expansion of European international society (O'Hagan 2002; Keal 2003; Suganami 2003; Suzuki 2005), the historical focus of scholarship within the English School tradition has overwhelmingly been on the story of how non-Europeans interacted under the context of European domination. However, the interactions of the European and non-European actors are naturally much more multifaceted. This chapter attempts to address this lacuna by examining interactions between the Netherlands and Japan during the Tokugawa period. It argues that the Dutch played a crucial role in the construction of the Japan-centric tribute system within the East Asian international order. It also seeks to rethink the longstanding English School question of whether in the absence of a common civilization two different polities were able to construct some form of meaningful social life between one another.

Rethinking European and non-European relations before European expansion

The most important work on European and non-European interactions from the English School is undoubtedly *The Expansion of International Society*, edited by Hedley Bull and Adam Watson. As can be gleaned from the title, the empirical focus is very much on a story of assimilation, when the non-European polities were subsumed into the European-dominated international order. The story here is (as noted above) presented generally as a 'success story': as English School scholars viewed (European) international society as a positive force that would mitigate the Hobbesian insecurities generated by anarchy, its expansion was seen as a desirable development. This also meant that the English School scholars' view of the history of European international relations was highly one-sided, where 'Western societies' achievement of religious and political tolerance originated with Westphalia and was furthered by subsequent treaties and conventions while non-Western societies' lack of religious and political tolerance was shaped by their intolerant and despotic past' (Kayaoglu 2010: 195). Furthermore, because of the agenda to demonstrate that a common civilization was not a prerequisite to the spread of global norms, English School scholars depicted this historical development as a relatively linear process. As O'Hagan notes, '[u]niversalist overtones surface in Bull's discussion of international society … he does imply that development and progress have been linked to the expansion of the European system' (O'Hagan 2002: 129). As Adam Watson (1984a: 31) notes:

> … the nineteenth century is notable for the creation throughout Asia, Africa, and Oceania of Europeanized or Westernized élites. The Europeans and the Americans offered the instruction, and usually met with an enthusiastic response … The mastery of Western governmental practice and military technology enabled these élites to run a modern state.

It is important to note, however, that this narrative effectively robs many of the non-European polities and their peoples of their agency. Because of the English School's implicit assumption that European institutions and norms are inherently progressive, the adoption of European institutions and norms was seen as something rational and inevitable – provided that the non-Europeans desired to escape from their 'backward' or 'regressive' state of living (Suzuki 2009: 17–25). What is also visible here is that Watson's narrative – alongside other authors of the English School – at times appears almost blithely oblivious to the fact that the expansion of European international society was often the result of outright European coercion. Suganami (2003: 263) makes this point nicely with reference to Watson's discussions of the expansion of European international society when he notes:

> What is disturbing here is the conspicuous absence of the story-teller's aside, reminding the reader of one key feature of the [nineteenth] century – imperialism. Watson's observation … describes the state of the mind of the complacent and ill-informed 19th-century European; the transformation could not have seemed so utterly innocuous to others.

This of course does not imply that the English School has been oblivious to European imperialism and its evils. Indeed, scholars working in this tradition have come a long way in terms of critiquing the Eurocentricity of earlier works. Edward Keene, for instance, has provided a detailed critique of the English School's one-sided interpretations of Grotius and identified a more coercive mode of interaction adopted by European international society in its interactions with those polities and peoples deemed 'uncivilized' (Keene 2002). Paul Keal (1995, 2003; cf. Anghie 2006) has similarly focused on the plight of indigenous peoples in the wake of the expansion of the society, and has exposed how even natural law, which ostensibly entitled all of humankind its protection, could function to justify the conquest of indigenous peoples and the extinguishing of their cultures.

There does remain, however, an element of Eurocentricity in these arguments, albeit implicitly. While the new approaches adopted by the more recent English School works *have* succeeded in puncturing the more unbridled celebration of Europe's 'civilizing' mission that accompanied the expansion of international society, their empirical focus is still firmly on the expansion of *Europe*, with European states the subjects. John M. Hobson (2007a: 93) has called this particular narrative 'subliminal Eurocentrism' which is critical of the West but nevertheless assumes 'that the West lies at the centre of all things in the world and that the West self-generates through its own endogenous "logic of immanence", before projecting its global will-to-power outwards through a one-way diffusionism'. The history of Europe's diplomatic interactions with the non-European world *prior* to the expansion of European international society, which was often dictated on the latter's terms, is relegated to the realm of insignificance.

Two factors account for this. The first is related to the construction of a European identity based on exceptionalism, particularly in the case of authors such as Bull or Watson. As Kayaoglu has pointed out, this 'European self-identification depended on various European other-identifications' based on the latter's weakness and 'backwardness'. The story of non-European strength – both militarily and politically – is hardly conducive to constructing this ethnocentric identity, particularly as 'the assertion of the complete superiority and exceptionalism of the European political and legal order … necessitated the European willingness to spread it' (Kayaoglu 2010: 206).

The second factor is the English School's distinction between an 'international system' and an 'international society'. Hedley Bull and Adam Watson (1984b: 1) have defined an international society as:

> … a group of states (or, more generally, a group of independent political communities) which not merely form a system, in the sense that the behaviour of each is a necessary factor in the calculations of the others, but also have established by dialogue and consent common rules and institutions for the conduct of their relations, and recognize their common interest in maintaining these arrangements.

As international society is only presumed to have attained its global reach by the end of the nineteenth century, any interactions between European and non-European polities are implicitly regarded as 'systemic'. This implies that no common rules and interests emerged between the two, and it is therefore less worthy of scholarly attention. This is problematic for a number of reasons. From a more sociological point of view, it is wrong to conclude that common norms and interests could not emerge between the Europeans and non-Europeans until European international society expanded and the Europeans imposed their own rules on the rest of the world. Rudimentary, functional agreements of cooperation between polities can develop into deeper societal relations (Buzan 1993). Furthermore, even if we do accept that it is extremely difficult to establish some objective 'tipping point' when 'systemic' relations become 'societal', meaningful interaction between two polities is extremely difficult without some societal relations emerging at some point (James 1993: 273).

In addition, we should also note that there is thus no reason to assume that non-European polities' relations with the Europeans *prior* to the expansion of European international society were somehow uniquely 'systemic' and vice versa. Many non-European post-colonial states still harbour deep-seated resentment and suspicion towards continued Western domination in international society today. Non-European states' winning of independence and subsequent entry into the society was based on their utilization of inherently European norms such as sovereignty or national self-determination and this is often (rather complacently) accepted as evidence of a deep acceptance and strong identification with the Western-originated international order today (Reus-Smit 2002: 503). However, such perspectives fail to reflect on whether 'the rules and institutions of international society represent a genuine pluralist consensus or an expression of cultural hegemony' (O'Hagan 2005: 217; cf. Brown 1995: 191). It is important to keep in mind that this fact may not necessarily reflect an acceptance and internalization of Western norms, but rather because in the face of continuing Western domination, non-European peoples had no other choice (see Bull 1984a).

Japan and Europe in the East Asian international order[1]

The aforementioned shortcomings in conventional English School accounts of the world *before* the expansion of European international society are brought into sharper focus by the case of Japan's foreign policy prior to 1853. The most important English School account of this particular period is of course Hidemi Suganami's contribution to *The Expansion of International Society.* Suganami's rendition of Japan's international relations prior to the arrival of European international society mirrors Eurocentric Japanese history which was heavily influenced by modernization theory. According to this interpretation, Japan's encounter with European international society heralds the beginning of a (felicitous) journey from feudalism to modernity (Dower 1975: 3–101). Therefore, as Tashiro (1982: 283) notes, Japan's policy of refusing contact with the European world except the Netherlands evokes an image of 'a closed society having absolutely no contact with any other nation' and 'we find the Edo period [the Tokugawa period] itself being defined as "the period of isolation" or "the dark ages" or "the beginning of the Japanese tragedy."' While this approach can be criticized for supplying a rather sanitized story of European imperialism, what is of most interest to our discussion is how the story of interactions between non-Europeans fades into the background. The most 'significant' historical event worth of scholarly scrutiny in the diplomatic history of East Asia is seen as the events surrounding the 'impact of the West'.

With regard to any interactions *prior* to Japan's 'entry' into European international society, Suganami unsurprisingly provides a relatively sparse narrative, pointing out that Tokugawa Japan had effectively cut itself off from interacting with any European powers with the exception of the Netherlands, whose merchants were required to reside in a secluded artificial island in Nagasaki called *Dejima*. This policy – implemented gradually between 1633–41, and known as *sakoku* – is implicitly regarded as creating 'systemic' relationships between Japan and European international society, as both sides merely 'accept[ed] the empirical reality of each other's existence' and had yet to take 'the additional step of respecting one another's right to sovereign independence' (indeed, such a concept did not enter the Japanese lexicon until the nineteenth century) (Linklater and Suganami 2006: 130). It is only when Japan began to enter into relations based on international treaties and law in the late nineteenth century – unequal though they were for the Japanese, who were branded 'semi-civilized' – that its relations with the Europeans are said to have become societal in nature.

It would of course be unfair to criticize Suganami for not providing a more detailed analysis of the relations between Tokugawa Japan and the Europeans, as his explicit goal was to explore the process by which the Japanese were incorporated into European international society. Yet, the relative silence accorded to the relations between the Netherlands and Japan *prior* to the expansion of the society in itself sheds light on the enduring legacy of Eurocentrism. As with other studies of the English School, the fact that Japan had limited its interactions with Europe is deemed as 'isolation' from the *world*, which ignores the diverse diplomatic relations the Tokugawa shogunate maintained with Korea, China and the Ryūkyū kingdom (Toby 1991: 4–6). This notion – which only sees diplomatic interactions dictated by Europeans and passive non-Europeans as worthy of scholarly attention – is further strengthened when we note that *societal* relations are only deemed to have emerged between the Japanese and Europeans in the late nineteenth century, once Japan had been forcibly assimilated into European international society. Any possibility of deep, societal (and therefore 'progressive') relations emerging between Asia and the European world on *Asian terms* is never seriously considered. This could not be further from the truth, however: while many conventional historical studies (on which Suganami's work draws) tended to concentrate heavily on examining Japan's relations with the Netherlands, Japan was at this time in a position to impose its own norms of diplomatic conduct to 'foreign' polities, and this would mean that the Netherlands would be incorporated into the social structures of a particular international order based on the terms of the Japanese.

The East Asian international order and Japan

Japan's international relations under the Tokugawa shogunate were governed by the constitutional structures of the East Asian international order, which had originated in China and were undergirded by Confucian ideology (Reus-Smit 1999; Fairbank 1968; Zhang 2001: 56; Mancall 1984: 13–39). As Yong-jin Zhang (2001) has noted, in this social setting interstate relations were conducted on a hierarchical basis, rather than sovereign equality: there was a 'centre' that stood at the apex of this order – usually China – and its ruler was assumed to be a mediator between humankind and heaven. The maintenance of these social hierarchies was seen as a crucial means of attaining social harmony and ultimately the happiness of the peoples of the respective member states of this international social order. Diplomatic relations between states operated on the basis of 'ritual justice', which meant that legitimate conduct was governed by elaborate rituals and ceremonies which functioned to maintain and strengthen the given social hierarchy of the East Asian international order.

These constitutional structures were expressed and reproduced through an institution known as the tribute system. Under this arrangement, diplomatic missions consisted of foreign emissaries presenting 'tribute' – gifts for the ruler of the state that occupied a higher place in the social hierarchy – in return for 'favours' such as investiture, trading rights and gifts of higher value. These diplomatic interactions were governed by a set of elaborate rules that served to illuminate that the ruler of the host country had attained the moral authority as an 'ethical ruler', to the extent that foreigners were presenting tribute and placing themselves *below* the ruler in the Confucian social hierarchy. This made it a powerful means by which the rulers of polities within the East Asian international order legitimated their rule.

Japan's longstanding historical legacy of interactions with China and Korea (both key members of the East Asian international order) and cultural learning from the Asian continent meant that the Japanese were deeply embedded in the social structures of this Confucian-influenced international social environment (see Howell 1998: 119–20). This meant that many of their diplomatic institutions were broadly similar to those of other members of the East Asian international order. This even extended to Japan's so-called isolationist policy, the *sakoku*. Tanaka Takeo notes the remarkable similarity of Japan's *sakoku* policy with the 'seclusionist' policies (known as *haijin* 海禁 in Chinese) undertaken by the Ming and Korea. The policy consisted of the state banning its peoples from private overseas travel and trade. It was first implemented by the Ming Dynasty of China to prevent 'Japanese' piracy (known as *wakō* or *wokou* 倭寇) and monopolize trade (Tanaka 1975: 85–86). The policy eventually evolved to support the hierarchies of the East Asian international order and was adopted by China's neighbours. By forcing foreign merchants to participate in 'official' trade and its rituals, the member states attempted to demonstrate and shore up their 'superior', 'civilized' status. 'Therefore', Arano (1988: iv) argues, Japan's policy of *sakoku* 'was not necessarily a policy that "closed a state's borders" but a policy to recognize a given state's desired way to conduct diplomatic intercourse'.

However, by the time of the Tokugawa, Japan's acceptance of Sinocentrism was somewhat waning. Instead, the Japanese placed *Japan* at the apex of civilization.[2] This process was hastened by the collapse of the Ming and the rise of the 'barbarian' Manchurian Qing Dynasty in China. The Tokugawa shogunate refused to send any tribute missions to China and admit to the latter's 'superiority'. Instead, the Japanese began to invite their own neighbours to send tribute to Japan instead. These diplomatic policies were hardly 'isolationist': rather, as Toby (1984: 96) argues, they:

> … should be seen as part of a much larger foreign policy embracing all of Japan's world. That foreign policy made tradeoffs … The

> Tokugawa tradeoff was between unrestricted foreign commerce, on the one hand, and the demands of sovereignty, security, and legitimacy, on the other.

In order to reinforce the notion that the Tokugawa clan had unified Japan and were the rightful rulers of the land, the first *shōgun*, Tokugawa Ieyasu, appealed to the fact that he had 'unified the country … rectified administration and brought prosperity to the people; and his dynasty had already attained its third generation' (Toby 1984: 60), all of which fitted the Confucian normative ideal. Korean and Ryūkyū (present-day Okinawa) rulers were encouraged to send missions in order to demonstrate that the *shōgun*'s prestige had spread far and wide. As one Japanese feudal lord put it in the context of the Korean missions' visit to first Tokugawa *shōgun* Ieyasu's shrine in 1637, 'The Three Ambassadors paid their respects … solely because the three generations of peace [between Japan and Korea] and the peace of Korea are due entirely to the high grace of Tōshōgū' (Toby 1984: 99). The missions were hosted at the expense of the shogunate, a practice that mirrored the Chinese custom of paying for the expenses of the foreign emissaries.

Red-haired barbarians and the Japanese centre

What role, then, did the 'red-haired barbarians' – as the Europeans were called – play in Japan's international relations during this time? Relations between the Europeans and the Japanese were primarily based on trade. However, once Japan's feudal rulers had consolidated their control over Japanese territory, the Europeans could become useful tools by which to demonstrate the widespread nature of their prestige as righteous rulers and consolidate their legitimacy to rule. To this end, the Europeans were incorporated into the Japan-centric East Asian international order and interacted on the terms dictated by the Japanese, playing a crucial role in the maintenance of its structures.

Of course, societal relations between these two polities were not based on international law or natural law as European international society was, given that the Europeans and the Japanese did not share the same philosophical traditions or god. Instead, in accordance with the principle of hierarchical sovereignty, European states that interacted with members of the East Asian international order had to be assigned a place within the social/civilizational hierarchy. This was common practice for other states in East Asia, most famously China, and the Europeans who wished to cultivate some form of relations – trading or political – had to conform to these norms. Evidence of this can be seen from the fact that Toyotomi Hideyoshi (豊臣秀吉), the feudal ruler of Japan in 1590–98, had ordered the Portuguese governor-general of India and the Spanish rulers of the Philippines in 1591 to present tribute. The Tokugawa shogunate was no different. As with the Chinese world order, '[t]he Tokugawa regime created for itself a naturalized version of the sinocentric world order of a civilized core surrounded by barbarian or at best imperfectly civilized peripheries' (Howell 1998: 119). By the time of the seventeenth century, this had evolved into a division of the world into Japan (*honchō* 本朝), China (*kara* 唐) and the Occident (*seiyō* 西洋, which encompassed the 'outer barbarians'), primarily based on civilization. While the former were characterized by shared 'writing systems and ethics [i.e. Confucian culture]', the latter were 'countries that used horizontal writing systems, unlike Chinese (唐土と差ひて皆横文字の国也)' (Tsukamoto 1979: 9). Under this scheme, Japan was placed within the Chinese cultural world (albeit at the apex), which the Japanese astronomer Nishikawa Joken (西川如見) labelled *gaikoku* in his 1708 work *Ka i tsūshō kō* (華夷通商考). Meanwhile, the Dutch were placed within the sphere of the *gai i* (外夷), or 'outer barbarians'. Similar schemes were deployed in the encyclopaedia *Wakan sansai zue* (和漢三才図会), which was published in the early eighteenth century. The book placed the Dutch and the Spanish in the category of the *gai i* alongside three-bodied monsters, implying that the Europeans were entities only '"one class above beasts" that did not meet the civilizational standards of the Chinese cultural world' (Tsukamoto 1979: 10; Tashiro 1982: 289).

It is worth mentioning, however, that the Japanese neither entirely saw their civilization as '*progress* in relation to the political, economic, and social institutions and practices of a society', nor as 'a *single and universal* concept used to describe a teleological *process* through which individuals and groups became civilized' like their Chinese or nineteenth-century European counterparts (Hirono 2008: 22). Whereas the Chinese (who shared with the Japanese broadly similar culturalist views of the non-Chinese world) tended to hold that Chinese civilization was inherently superior and that the Chinese emperor could guide the 'barbarians' to progress, the Japanese came closer to what Miwa Hirono (2008: 23; cf. Huntington 1996) calls a 'pluralist conception of "civilizations"', and viewed these three worlds as 'self-governing worlds with their own states, societies and cultures' (Arano 1988: 53). This divergence, Arano (1988: 53) notes, possibly originated from the fact that 'the Japanese state authorities had no choice but to construct their self identity [自己 *jiko*] under the political and cultural influence of China, India or Korea'. Nevertheless, the norms of the East Asian international order were a 'heavy chain that bound the inner and outer worlds of the early modern Japanese' (Arano 1988: 61), and the Tokugawa shogunate did insist that appropriate social rituals governing diplomatic conduct be observed by the 'outer barbarians'. As Tashiro Kazui notes, 'in [Japan's] dealings with the Ryūkyūs and Holland, which can hardly be described as normal diplomatic relations between countries of equal standing, Japan insisted that it be paid a sort of tributary mission' (Tashiro 1982: 289).

This, of course, does not imply that the norms governing 'ritual justice' in the Japan-centric tribute system had already been established by the Tokugawa shogunate when it began interacting with the Europeans. The rituals themselves undoubtedly bore the hallmarks of the East Asian international order: in 1613 the English East India Company commander John Saris had to spend five days ‘to prepare and transport presents from the king to the Japanese shogun’, even before negotiations for the granting of trade privileges began. This practice was carried out in China as well, where some form of ‘tribute’ was required in return for benevolent ‘favours’ to be bestowed upon the European ‘barbarians’ (Klekar 2006: 93). Europeans – be they Portuguese, or Dutch – were often not spoken to by the *shōgun*. Instead, they were merely invited to display their gifts in front of the Japanese ruler, who would acknowledge their ‘tribute’ with a silent nod. A Dutch emissary to China 1667 was granted an audience with the Kangxi emperor in a style that also mirrored that of Japan, making this ‘style of audience a uniquely East Asian one’ (Nagazumi 1990: 99). Nevertheless, the Tokugawa shogunate did face the same problem that the Europeans encountered later when dealing with non-European peoples who had yet to be ‘digested’ (to use Wight’s words) into European international society (Wight 1992: 50). Indeed, at times ritual protocol was not always observed by the Japanese rulers, especially when the *shōgun* himself was either personally favourably disposed towards the Europeans or had specific desires for trading relations.[3] Nagazumi (1990: 96) also points out that in the early years of the rule of the first *shōgun*, Tokugawa Ieyasu (徳川家康), the historical records list Japanese, Portuguese and Spanish gifts to the *shōgun* together, which is indicative that ‘there were no established rituals for receiving gifts, including those from foreigners’. Consequently, a Portuguese mission sent by the governor of Goa (described as an *embaxadors* by the Dutch) in 1611 was provided with horses by the shogunate free of charge, and proceeded to the capital Edo (present-day Tokyo) dressed in elaborate velvet clothing and playing musical instruments. This, Nagazumi (1990: 99) argues, demonstrates that the Portuguese mission was treated in a similar style to a Korean tribute mission, and is indicative of the still somewhat fluid position of the Europeans in the hierarchical ordering of foreigners in Japan’s tribute system.

The audience question of 1627: consolidating hierarchical orders

The East Asian international order, however, ‘did not exist by itself; rather, it had to be reconfirmed in the arena of external relations’ (Arano 1988: x). As the Tokugawa shogunate sought to consolidate and legitimate its rule both to its domestic and international audiences, it soon began to formalize the institutional arrangements surrounding Japan’s interactions with the ‘outer barbarians’ and differentiate them from those who came from the East Asian cultural sphere. In this sense, the Europeans would serve as an important ‘Other’ that would function to highlight Japan’s own identity as a ‘civilized’ and paternalistically ‘benevolent’ head of the East Asian international order (cf. Neumann 1999).

A symbolic diplomatic incident that accelerated this process was the debate surrounding the visit of Pieter Nuyts, the emissary of the Dutch governor of Batavia, in 1627. Nuyts’s visit itself was prompted by a dispute between the Dutch and the Japanese resulting from the former’s establishment of a port in

Taiwan. Keen to set up a base for trading with China, the Netherlands had constructed a fortress in Anping, close to Tainan. In an attempt to improve the competitiveness of their trading vessels, the Dutch began imposing a tax on any exports from the port and sought to monopolize trade originating from Anping, bringing them into conflict with Japanese licensed trading ships. The Japanese side retaliated by closing the Dutch trading house (then still located in Hirado 平戸) and banning the Netherlands from trading with Japan. Nuyts was charged with the task of explaining the Dutch position and avoiding an escalation of this diplomatic row.

Nuyts arrived in Hirado in 1627, and immediately got entangled in a lengthy debate over his official status and the appropriate diplomatic rituals that should govern his interactions with the Tokugawa officials and the *shōgun*. The Hirado fiefdom officials, upon hearing that this emissary was an 'ambassador', proposed that Nuyts 'make the procession up to Edo as grand as he could' (Nagazumi 1990: 107), and word also came from the shogunate in Edo that it would pay for the horses and porters needed for the procession. Here, we can see that the Dutch mission was treated as equivalent to a tribute mission from a member of the East Asian international order, such as Korea or the Ryūkyū kingdom. However, the Japanese officials' interpretations of the nature of Nuyts's mission, as well as its very compatibility with the normative framework of the East Asian international order, was soon put to the test when the Japanese found out that Nuyts had a number of diplomatic matters to discuss with the *shōgun*. According to Japan's diplomatic protocol, all visits by the Dutch 'were to pay a visit to the *shōgun* and present him with gifts. If there were any other matters to discuss, these were to be requested via appropriate intermediaries, and to be dealt with by the appropriate offices', rather than through direct negotiations with the *shōgun* himself (Katō 1981: 84).

Nuyts's own official position also came under scrutiny during this time. Most of his stay in Edo, notes Katō (1981: 85), was spent dealing with repeated questions over whether the ambassador:

> (1) had come directly from the Netherlands; (2) was appointed and sent by the Dutch 'king'; (3) the document that accompanied him was written and signed in Holland; and (4) the Governor of Batavia was related to the 'king', or merely a subject.

Nuyts explained that although he was an ambassador of the governor of Batavia, 'since the Governor governed the Dutch East Indies on behalf of the king, he was effectively the sovereign of the territory, and it was *international custom* to treat such ambassadors [despatched by sovereigns] with courtesy', and on this basis 'he should be allowed an audience with the *shōgun* as soon as possible' (Katō 1981: 86, emphasis added). The Japanese officials were 'deeply disappointed' that Nuyts was not an ambassador from the king of Holland, particularly as they had given him the highest level of treatment in accordance with their diplomatic rituals, and Nuyts's diary records that the feudal lord of Hirado was almost ordered to commit suicide for his erroneous reporting of the nature of the Dutch mission (Nagazumi 1990: 110).

In the end, Nuyts was compelled to sign a statement that his mission was 'to give thanks to the favours given to Holland by the *shōgun*', and that he was not a special emissary from the Netherlands, and that his 'mission was not to make any demands to the *shōgun* with regard to the dispute in Taiwan' (Nagazumi 1990: 108–9). Although the officials of the shogunate attempted to alter Nuyts's status as a 'merchant' to allow an audience with the *shōgun* Tokugawa Iemitsu, ultimately the Dutch letter was deemed inappropriately discourteous for the Japanese ruler's eyes, and Nuyts was given the explanation that 'Ambassadors come to pay their respects to the ruler when they visit Japan, not make demands' (Nagazumi 1990: 110). At a certain level, we can see Tokugawa attempts to ensure that the *shōgun*'s image as a benevolent, paternal figure (something that fitted the Confucian ideal) was not threatened by an emissary from a lower-ranking polity making demands and implying that the special favours given by the ruler of Japan were somehow inadequate. Nuyts, of course, failed to see this point of view, which is unsurprising given that he was operating under European normative structures. He was forced to return, furious, to Taiwan empty handed. He meted out his revenge on the Japanese traders in Taiwan, only to be 'kidnapped by them and imprisoned in Japan, before the authorities at Batavia could secure his release by eating humble pie' (Boxer 1984: 539).

The debacle of the Nuyts mission prompted the new governor, Jacques Specx, to adopt a more conciliatory stance. He ordered the despatch of a new mission to Japan in 1630, but this time told the mission that the Dutch East India Company (VOC) 'should not seek empty glory and honour; instead, we need to ensure that we can obtain profits from free and unhindered trade with Japan as before, as well as let the Japanese enjoy benefits [from trade with the Netherlands]' (Katō 1981: 88–89). Crucially, Specx appears to have decided to accept the Japanese hierarchical ordering of Holland (if not sincerely), and warned the Dutch mission that 'even if they encounter Japanese arrogance to keep in mind that "the Japanese are a great and proud people and the Dutch are a small and petty people"'. This policy, Katō (1981: 89) argues, eventually evolved into the use of the word *keizers eigen volck* (the servants of his imperial majesty), which the Dutch would use when addressing the *shōgun*. Specx's tactics were ultimately to bear fruit, and the Dutch were allowed to reopen their trading house in Hirado and restart trade in 1633.

The Netherlands continued to play a crucial role in the construction of Japan's own 'tribute system' until the expansion of European international society into East Asia. Such handling of the Dutch served to confirm the 'outsider'/'outer barbarian' status of Holland in the Japan-centric tribute system. The Dutch were required to pay a visit to the *shōgun* in Edo (present-day Tokyo), and this occasion served a similar function to tribute missions by the Koreans and Ryūkyūans, reinforcing the social structures of the East Asian international order, albeit with Japan as the centre, rather than China.

By the time the Tokugawa shogunate had consolidated its rule, there was not much room for manoeuvre for the Dutch, as far as diplomatic rituals were concerned. There now existed strict, hierarchical 'diplomatic' protocols which governed the etiquette of the Dutch visit, which befitted their 'outer barbarian' status. The *shōgun* did not 'receive' the delegation as he did for the Koreans or the Ryūkyūan emissaries; rather, 'the shogun was recorded … as "viewing" the Hollanders, much as he had "viewed" *entertainments* offered by the Korean equestrian troop or the Ryūkyūan musicians' (Toby 1984: 190). The Dutch were not allowed to speak to the *shōgun*, and neither were they allowed to present him with any petitions. The differential treatment given to the Dutch by the Japanese (in contrast to their reception of the Korean or Ryūkyūan missions) mirrored the Canton system of trade adopted by Qing China, where 'outer barbarians' were segregated from the local populace and allowed limited trade as an 'imperial favour'.

The Dutch were well aware of their position within the Japan-centric tribute system, and employees of the Dutch East India Company resident in Nagasaki were instructed to be 'armed with modesty, humility, politeness and friendship, *always behaving as inferiors*' (Boxer 1984: 539, emphasis added). In a fascinating account that demonstrates how Tokugawa diplomatic symbols had evolved since the time of Tokugawa Ieyasu, Engelbert Kaempfer, the German physician to the Dutch delegation, recorded that in the audience with the *shōgun*, the Dutch trade representative greeted the Japanese rulers as follows: 'kneeling, he [the Dutch trade representative] bow'd his forehead quite down to the ground, and so crawl'd backwards like a crab, without uttering one single word. So mean and short a thing is the audience we have of this mighty Monarch' (cited in Toby 1984: 193). The Dutch were even prepared to play the fool in their audience with the *shōgun* if necessary. Kaempfer reported that:

> There was a Royal Viewing (*jōran*) of the four Hollanders [at which the Europeans were required] to walk, to stand still, to compliment each other, to dance, to jump, to play the drunkard, to speak broken Japanese, to read Dutch, to paint, to sing, to put our cloaks on and off … In this manner, and with innumerable such other apish tricks, we must suffer ourselves to contribute to the Emperor's and the Court's diversion.
>
> (Toby 1984: 194)

A number of points can be made from these incidents. First, we see that the shogunate's handling of the Nuyts affair demonstrates that by the time of Tokugawa Iemitsu, Japan had already established fairly concrete regulations concerning foreign audiences with the Japanese ruler according to the official status of the emissaries. Second, it is clear that the procedural norms governing diplomacy in Europe – such as the treatment of an ambassador – were deemed as simply incompatible with East Asian diplomatic norms and rejected by the Japanese. Nuyts's invoking of *international customs* meant very little to them, as Japan's own normative frameworks were governed by those of the East Asian international order, rather than European international society.

In contrast to the late nineteenth century, the Europeans were in no position whatsoever to force the Tokugawa shogunate to accept and participate in European norms of interstate interaction (see also Massarella 2001).[4] As far as the Dutch were concerned, their priorities lay in establishing themselves as trading partners of the Japanese. During the 1627 audience dispute, the Netherlands was still competing with the Portuguese for the *shōgun*'s trading privileges (their other rivals, the English, had closed down their trading house in Hirado in 1623, following poor profits). Given the shogunate's suspicion of the Portuguese Jesuits, the Dutch were well aware that if they refrained from proselytizing in Japan, they had the chance to monopolize Europe's trade with Japan – and they eventually succeeded in this endeavour in 1639, when the Portuguese were banned from sailing to Japan. After this date, it made even less commercial sense for the Dutch to get entangled in disputes with the Tokugawa shogunate over diplomatic rituals. As Grant K. Goodman (2000: 240, footnote 11) notes, '[u]p to the beginning of the eighteenth century the profit for the Hollanders on their annual trade with Japan was over 50 per cent, thus making it the richest Dutch trading post in the East'. Dutch trading fortunes waned later on, but even then there were opportunities for financial gain to be made by engaging in smuggling with the Japanese (Keene 1969: 6). Furthermore, there is some evidence of caution of the military prowess of the Japanese on the part of the Dutch. Referring to Holland's harassing of Portuguese ships, one VOC employee wrote to Governor of Batavia Jan Pieterszoon Coen that:

> … the Shogun of Japan was 'no king of Makassar' and he would not tolerate the Dutch trying to intercept Portuguese shipping in Japanese waters. He added that probably the king of Makassar did not like the Dutch violating his neutrality either; but that he lacked the force to stop it, whereas the Japanese most certainly did not.
>
> (cited in Boxer 1984: 539)

In light of these considerations, the Dutch were effectively willing to incorporate themselves into the East Asian international order for the sake of maintaining harmonious trading relations, and it is important to note that this arrangement lasted for more than 200 years – this is still longer than Japan's encounter with and subsequent entry into European international society, which was in its 157th year at the time of writing. They shied away from confronting Japanese officials over diplomatic protocol by simply not sending any ambassadors or letters from the Dutch sovereign (Nagazumi 1990: 125). Instead, they limited their interactions with Japan to a 'commercial' basis, with any diplomatic interactions taking place between the Dutch head of the trading house and appropriate officials of the Tokugawa shogunate.

Rethinking international order and European domination

What implications does the case of European–Japanese interactions prior to 1853 generate for us? First, the case of Japan's interactions with the Netherlands suggests that a common culture may not be a prerequisite for societal relations between two polities. Martin Wight was of course explicit on the need for cultural homogeneity, arguing that an international society 'will not come into being without a degree of cultural unity among its members' (Wight 1977: 33). A shared civilizational background in Christianity may have indeed played an important role in the emergence of European international society, and a shared Confucian civilization was also a crucial ingredient for the emergence of the East Asian international order.

Alexander Wendt, however, contests this view when he points out that the 'mistake made here is thinking that "culture" (shared knowledge) is the same thing as "society" (cooperation)'. However, '[s]hared knowledge and its various manifestations … are analytically neutral with respect to cooperation and conflict … there is nothing about the absence of shared knowledge, a world of only material forces, that necessarily implies a war of all against all' (Wendt 1999: 253–54). The case of Japan and the Netherlands seems to support this. The fact that the Dutch went along with playing their lowly social role in the Japanese tribute system also shows that both Holland and Japan recognized certain interests (if not common values) and 'regard[ed] themselves as bound by certain rules in their dealings with one another' (Bull 1995: 13), and this seems to suggest that the Japanese and Dutch had developed some form of *societal* relations as defined by Hedley Bull. In spite of the lack of a common cultural heritage, both states were able to maintain relatively peaceful relations with each other for more than 200 years.

Second, an examination of relations between Japan and the European world prior to the expansion of European international society seems to point to an enduring tendency towards promoting homogenization in the English School. The neglect of the history of interactions between the European world and the non-European world rests, as mentioned previously, on the artificial distinction between 'societal' and 'systemic' relations. Despite the empirical difficulties of distinguishing *when* and how 'societal' relations emerge between two actors, many studies of the English School have assumed that such relations were only possible when non-European polities were subsumed into European international society and adopted *European* norms values. As 'societal' relations are seen as inherently desirable, this has the unfortunate effect of reproducing Eurocentric notions of only seeing diplomatic relations with the *West* as meaningful and worthy of scholarly attention. In this sense, then, the 'system' and 'society' distinction serves as an intellectual device to 'cut off' an important legacy of the history of humankind into oblivion and moral 'backwardness'. Such views are no doubt strengthened by the English School scholars' own moral convictions that societal relations with the European powers had the positive effect of spreading progressive norms (associated with what Wight called 'Western values') throughout the world (Wight 1966; cf. O'Hagan 2002; Suganami 2003: 259–66), and we continue to see similar assumptions (albeit implicitly) in some liberal thinkers today (see Bowden 2004; Reus-Smit 2005).

Of course, it is hard to deny that many of the Dutch were participating in the reproduction of the social structures of the Japan-centric East Asian international order somewhat reluctantly for instrumental reasons. This was underscored in 1804, when Russian Navy Captain Adam Krusenstern, who had sailed to Japan to seek diplomatic and trading relations, commented that 'it is to be regretted that an enlightened European nation, owing its political existence to a love of freedom, and which has acquired celebrity by great actions, should so far debase itself from a desire of gain' (cited in Keene 1969: 5). The then Dutch trade representative Hendrik Doeff retorted that while Krusenstern may have found such rituals humiliating to European sensibilities, '[t]he compliments we observe with respect to the Japanese are the same as they themselves show each other … Besides, one cannot expect that a nation to which one comes in order to seek the friendship of the same should conform to the customs of the visitors' (cited in Keene 1969: 5, footnote). Later Dutch observers such as C.T. van Assendelft de Coningh would also write approvingly in 1851 (just two years before the arrival of European international society in Japan) that the Tokugawa shogunate's strict control of the foreigners was a wise one, 'for only in this way could peaceful relations have been maintained for so long'. De Coningh was also critical of 'the falsehoods spread abroad by foreigners in Dutch service', such as Kaempfer (who was German), '"who displayed the baseness of ascribing to us the same creeping role in front of the Japanese as they themselves showed in front of us when they crawled to us begging for a job"' (Paul 1977: 364). While we do need to take into account the possibility that both Doeff's and de Coningh's accounts may be somewhat patriotic, aimed at quelling European disapproval towards Dutch policy vis-à-vis Japan, it is nevertheless interesting and significant that even some Europeans did not always see the international normative structures imposed upon them by Japan as problematic, and had even begun to see some normative benefits of operating under the East Asian international order.

Conclusion

The examination of diplomatic relations between European and non-European polities *prior* to the expansion of European international society, while a neglected area of research within the English School, has the potential to provide us with some insights into international politics today.

Despite their increasing acknowledgement of the darker history of the expansion of European international society, it seems that many English School scholars – as well as scholars who advocated the global dissemination of liberal democratic governance in the context of the post-Cold War 'liberal triumphalism' – still adhere to the view that only cultural/civilizational homogeneity can achieve a truly peaceful, stable international order (Fukuyama 1989; Fidler 2000, 2001; Clark 2005: 180). This can be seen from the English School's neglect of the period when Europeans interacted with the non-European world on the latter's terms. This, to my mind, seems to mirror a tendency to obscure 'the inter-civilizational dialogues that played such an important role in the making of Western and global modernity' (Hobson 2007b: 417). The construction of this narrative in turn serves to preserve the Eurocentric myth of an uncontaminated 'Western' civilization which is 'a pure, self-constituting entity that created modernity through its own indigenous properties' (Hobson 2007b: 423). This seems unwittingly to reveal the deep-seated notion among English School scholars that European/Western civilization and political thought is an inherent force for good, and that contamination by non-European ideas and thoughts can only lead to moral regression. Such ethnocentric fears are today epitomized by the 'clash of civilizations' thesis forwarded by Samuel P. Huntington (1996), and also visible whenever a non-European state was seen on the ascendancy: in the late 1980s this took the form of the 'Japan threat' thesis, 'which drew on the "traditional" racial and cultural stereotypes to narrate Japan's threatening "otherness"' (Bukh 2010: 87; cf. Huntington 1993; McDowell 1990). We are beginning to see the same phenomenon with reference to China (Halper 2010).

More recently, however, the USA and its allies have been bogged down in a seemingly unwinnable war in Afghanistan and Iraq, with their hopes of constructing a liberal democracy in these lands seemingly in ruins. This has forced students of international relations to reflect on the limitations of imposing self-perceived 'progressive' norms on different cultures. This, of course, has long been the 'central dilemma for international society' and many scholars who work in this tradition (O'Hagan 2005: 223). While the discussions above may have served to highlight his more Eurocentric tendencies, it is important to acknowledge that Hedley Bull himself oscillated between his ideological commitment to Western values and recognition that in a much more culturally diverse international society, different cultural values would have to be incorporated (Bull 1995: 305; Dunne 1998: 148).

The historical record presented here and in other chapters of this volume may give us cause for some optimism, however. The notion that different cultures could easily clash has been criticized by Jacinta O'Hagan (1995: 24–25), who has pointed out in her critique of Huntington's 'clash of civilizations' thesis that coexistence was possible between two different civilizations. Very little research has been carried out on how different polities may be able to establish societal relations without reverting to coercive 'civilizing missions' often resented for the almost inevitable baggage of cultural superiority.[5] One of the keys to answering this question seems to lie in the examination of the history of European and non-European interactions *prior* to the expansion of European international society, and it is high time that that English School took up this challenge.

Notes

1 This section has been published in Suzuki, S. (2012) 'Europe and Japan', *Review of European Studies* 4(3): 54–63, particularly pp. 55–59, and has been revised and reproduced here with permission.

2 It should be noted, however, that Japan was not exactly alone in its resistance towards Sinocentrism: states such as Korea and Vietnam defied Chinese claims of superiority at various points in their history. This, of course, has interesting implications for recent arguments (Kang 2007) that claim an Asian tendency to accept hierarchical international relations.

3 The first *shōgun* Tokugawa Ieyasu was particularly known for his openness towards Europeans, even appointing William Adams, a Briton, as his adviser. A delegation from the Netherlands (1611) and the former Spanish governor of the Philippines (1609) were both spoken to by Tokugawa Ieyasu during their respective audiences with the *shōgun*. Nagazumi (1990: 100–1) states that this was possible because of Adams' personal introduction of the Dutch, as well as the *shōgun*'s own desire for trade with Spain.

4 This was the case for the English as well. The attack on Chinese, Portuguese and Spanish ships by the joint fleet of English and Dutch East India Companies in 1621 near Japanese waters incurred the displeasure of the Tokugawa shogunate, which imposed further restrictions on English and Dutch trading privileges. John Osterwick, based at the English trading house in Hirado, wrote at the time that 'the pres[en]t estate of this Countrie, towching our selfe & the Hollanders, you may please to vnderstand that wee stand vppon ticklish points of loosing our priviledges & to be put out of the Countrie' (cited in Massarella 2001: 26). This again is indicative that the Europeans were in no position whatsoever to dictate the terms of how they interacted with the Japanese.

5 To my mind, even pluralists may not have the advantage here, as they also require a certain degree of homogeneous political culture – predicated on the acceptance of the *Western* institution of the sovereign state and other institutions of international society – and do not allow for genuine tolerance of different cultures and arrangements of international life.

5 A corrupt international society

How Britain was duped into its first Indian conquest

Darshan Vigneswaran

… you will find how you became engaged in the founding and erecting a new Fort.
(Company servant preparing London for the news that they had become a landed power in India)

Introduction

This chapter asks to what extent the English School concept of the 'expansion of an European society of states' helps us to understand early encounters between Europeans and non-Europeans on the Indian sub-continent. The chapter tests this framework through an in-depth analysis of official correspondence amongst members of the British East India Company at the founding moment of British sovereign authority on the Indian coast in the middle of the sixteenth century. The treaties established by the East India Company to establish forts at places like Armagon and Chennai constitute a key test case for the English School assumption that shared culture is a fundamental prerequisite of international sociability. East India Company servants and the negotiators for sub-continental princes and emperors came from different cultural universes. Would cultural differences prevent them from establishing agreements over who could rule what territory?

An English School approach provides us with few resources for understanding this important historical moment, because it fundamentally misrepresents the thoughts of the primary actors in this historical drama. While Company servants were interested in 'expanding' British territory, they had no interest in the expansion of any civilizational standard of rule. While some of the actors in these negotiations had European origins, they did not draw on any discernible 'European' norms to determine with whom they would negotiate, or on what terms. While normative concerns helped to shape how Company servants understood their role, these were consistently trumped by base material considerations. Finally, there was no actor resembling a 'state' participating on either side of these negotiations.

Having established this critical distance from an English School perspective, the chapter responds to the question framed by this volume's editors: can shared 'international' norms emerge in the absence of a common culture? Here, the chapter is more constructive while remaining sceptical of any hope for a resuscitation of an English School agenda. I argue that the lack of a common 'culture' between British agents and their sub-continental counterparts presented no significant obstacle to their efforts to realize shared norms for how territorial jurisdiction should be distributed. However, the set of norms that British agents and their sub-continental counterparts shared are not those that have been emphasized by English School theory: war, great powers, diplomacy, balance of power, international law or sovereignty. Rather, the contemporary term that best defines the shared normative orientations of these two sets of parties is 'corruption': a shared belief in their mutual prerogatives to profit personally from sovereign institutions and infrastructure, and particularly fortifications. Put simply, since both Company servants and the underlings of sub-continental lords came from parasitic, rent-seeking classes in their respective countries, both were keen to fabricate political institutions that ensured their continued mutual gain.

Understanding this phenomenon requires that we abandon misleading theoretical starting points like a hypothetical 'society of states' and focus our attention on the groups and individuals that actually determined when and where the British acquired fortified territory in India. The 'society' that defined, negotiated and contested the initial conquest of India, setting the terms for British expansion over the next two centuries, is perhaps best seen as a transnational clique of contending but cooperating agents, who sought to capitalize on the benefits of new international trade.

International Relations (IR) scholars commonly seek to use the empirical invalidation of one theoretical paradigm to support or buttress another. Arguments invalidating the power of ideas or norms invariably seek to advance the case for materialist or rationalist frameworks. This case study does not suggest that such an approach would be particularly helpful. Despite the fact that base material interests and the rational interests of traders help to account for much of the behaviour of inter-continental negotiators in this case study, this does not validate the merits of a materialist or rationalist approach. Indeed, the chapter suggests that such approaches do similar injustices to history. In particular, they fail to see that the 'corrupt culture' that undergirded the emergence of a rudimentary international society between the British and Indian rulers and allowed this transnational elite to divert public institutions towards private ends, was not a mere *post hoc* justification of utilitarian interests but a specific way of structuring how interests were perceived and calculated that was peculiar to a specific historical setting. As socio-cultural changes in the United Kingdom began to undermine the culture of 'old corruption' in the middle of the eighteenth century, the ability of British and Indian agents to collude in the expansion of British territory also began to give way. Hence, the rather ahistorical models of material forces or rational interests do not help us to understand why the terms of European expansion became more normatively charged over time.

The remainder of this chapter will examine how the servants of the British East India Company initiated the process of territorial acquisition on the Coromandel Coast of the Indian sub-continent during the 1620s–40s. I will build this argument by first describing the common approach of British and Indian rulers towards commerce in the sixteenth and seventeenth centuries. I aim to show that rulers on the sub-continent and in the British Isles regarded the exploitation of virgin markets as a process to be sanctioned and 'farmed' for personal gain. This account helps to contextualize a more specific historical case study, which shows how the British East India Company's employees in India decided to establish the first self-governed British settlements in India. That section shows how, despite the lack of direct approval from London, sub-continental Company employees were able, with the encouragement of local rulers, to establish a fortified presence at Armagon and Madras. In addition to the theoretical argument, this discussion has the added value of an empirical contribution to historical discussions – firmly situating Armagon [rather than Madras] as the site where British territorial rule on the sub-continent began. The discussion then goes on to show how these initial agreements set the pattern for British acquisitions over the next century until cultural changes in England set Indo-European relationships on a new path.

In constructing this argument, this chapter relies heavily on the archives of the East India Company, and specifically the 'Official Correspondence' (OC) records held at the British Library. These records consist of letters, instructions, memoranda, treaties and agreements sent amongst factories in Asia, and between factors (employees) and the directors in London. The collection provides insights into the activities, perceptions and rationales of the mid- and upper-level Company employees on the questions of expansion. These records, however, must be read as the product of a limited perspective, leaving out European competitors, junior factors and servants within the Company and indigenous rulers. This latter category is of significant concern for the argument of this chapter. The lack of first-hand sources on the preferences of Indian elites means that we must rely on factors' interpretations of their interlocutors' underlying motivations and culture.[1] This is complicated by the fact that the Official Correspondence documents are necessarily *political* tools, crafted by Company officials to convince their compatriots (and more importantly, superiors and shareholders) of the merits of a particular course of action. While these limitations are somewhat balanced by the fact that much of the correspondence reflects the product of collective decisions and/or meetings amongst various factors and/or ship captains, it is highly likely that much detail and nuance has been deliberately omitted or edited in the self-interest of the authors, requiring the analyst to rest significantly on interpretation and judgement.

Negotiating trading rights in seventeenth-century India

When British traders first began to arrive on sub-continental shores in the late sixteenth century, they were attempting to establish themselves in a system of naval trade between Europe and Asia that had been forged by the Portuguese, but was increasingly being dominated by the Dutch and the French. While European powers had established themselves as highly influential naval powers in the Indian Ocean, this did not translate into a capacity to dictate terms to local rulers. Instead, indigenous elites often skilfully played European powers against one another as the latter sought to learn how questions of power and privilege on the sub-continent could be decided, and to refine their approach accordingly. For the most part, this precluded the possibility of exercising any substantial forms of territorial rule. Europeans traded out of a series of 'factories' located in towns and markets at the discretion of, and under the control of, local rulers or fief-holders. Previous powers had pioneered the concept of a fortified settlement, but the primary purpose was the fortification of port towns against naval attack. As noted by I. Bruce Watson (1980: 71), '[w]hile the Portuguese concentrated on levying tribute from the indigenous maritime trade of the Indian Ocean, the Dutch utilized their forts to establish control over the supply of fine spices in the south-east Asian Archipelago'.

At the beginning of the seventeenth century, Queen Elizabeth awarded a Charter to a group of merchants to form the East India Company, in order that she might profit from the growing trade in the East Indies. For the first half of the seventeenth century the representatives of the East India Company were relative novices in this world and were forced to learn, mostly through experience, mimicry, and trial and error, how to establish their Masters' mercantile interests in the region. They came as supplicants to Indian courts, seeking trading concessions from indigenous sovereigns while competing with Dutch and Portuguese competitors for favour. Over its first two decades, the Company established a small collection of factories across India (Surat, Pettapoli and Masulipatam), Persia (Gombroon) and the East Indies (Bantam, Celebes), and also had establishments at Dutch and Portuguese settlements in St Tomé and Pulicatt. However, the agreements that the Company factors reached with local rulers to acquire fortified footholds on the sub-continent soon emerged as a key issue. Although piecemeal at first and heavily subject to the continuation of sound relations with inland powers, the coastline acquisitions of Madras, Bombay and Calcutta were key, becoming the cornerstones of British claims over Indian territory and setting the terms for subsequent expansion. The early Company utilized these outposts and the advantages of bullion exports to gain a foothold in a range of pre-existing Asian trading networks, acquiring calicoes, indigo, silk and pepper from a set of existing coastal entrepôts (Bassett 1998: 3).[2]

By the time the Company wound up its operations in the middle of the nineteenth century, Britain was the single greatest power on the Indian sub-continent, but the basic principles that undergirded these original agreements continued to define the constitutional character of their authority there. Key principles, such as the notion that British rulers would exercise ruling prerogatives on behalf of local rulers and that British laws would govern subjects of the British Crown emerged in these early documents and remained central to subsequent treaties. As a result, the East India Company records for this period provide us with unique insights into the process whereby the British participated in, internalized the principles of, and then subsequently realized the limitations of the 'international society' that prevailed in India during the seventeenth and eighteenth centuries.

The non-Eurocentric international order

In order to comprehend the character of this international society we need to engage critically with the Eurocentrism implicit in the English School's account of the expansion of international society. Most English School accounts implicitly assume that due to the socializing power of culture or the progressive nature of 'Western norms', many world regions will share norms of political authority (O'Hagan 2002; Suganami 2003). Rulers who originate from places that are far apart will possess different and/or incompatible social norms, while those who share neighbouring geographical origins will share certain understandings and norms about politics. In terms of English School theory, this assumption is embedded in the distinction between international systems as sites of minimal/mediated interaction and lack of social norms, and international societies as sites of dense/close interaction and commitment to common principles. This assumption is most clearly evident in work on the expansion of European international society and investigation of the deployment of a standard of 'civilization' in determining which non-European powers could exercise sovereign territorial prerogatives (Gong 1984a; Bull and Watson 1984a; Keene 2002; Suzuki 2009). These works frame the question of international order as one of resolving *differences* between European and non-European international actors, usually in favour of the former.

According to most English School accounts, one of the core principles of the European society of states is that of mutual recognition of exclusive sovereign authority as the principal norm of territorial rule. In this respect, the Peace of Westphalia in 1648 has long been regarded as the point when Europe shifted decisively to a norm of territorial sovereignty, meaning a single actor's exclusive possession of rights of rule over bounded and contiguous territory. This norm is said to have replaced the parcellized, multi-cephalic and blood-defined norms of political authority that prevailed during the preceding feudal era.

This Westphalian threshold has been criticized by several authors. Benno Teschke (2003) has argued, taking cues from Stephen Krasner's (1999) earlier suggestions, that Europeans did not decisively enshrine any specific form of political authority in the treaties of Osnabrück and Münster. Deploying a Marxist framework, he shows that the transition from feudal to modern sovereign forms of rule was not a Europe-wide process. Instead, this transition was instigated by the advent of capitalist agriculture and parliamentary rule in Britain. While capitalism created an important condition of possibility for an 'international society' based upon the principle of sovereign territoriality, this process was achieved unevenly in Europe and throughout the rest of the world.

The upshot of this argument is that Europeans did not share norms of territorial authority, even after 1648. In fact, since capitalism did not arrive in all parts of the continent all at once, but in England first and other parts of Europe thereafter, Europeans actually began to look more different than one another as the diachronous logics of capitalism and feudalism compelled their respective monarchs and rulers to relate to territory in different ways. For this reason, it makes little sense to approach early contacts between European and non-European rulers as the early efforts to iron out differences between their respective, regionally specific norms of territorial authority. To simplify, Europeans possessed multiple, and not singular, territorial norms when they arrived in India.

With this artificial rendering of difference out of the way, this chapter attempts to look at a more pressing question related to the early British conquests on the Indian sub-continent: what were the points of commonality between members of European and non-European international societies that created the rudiments of agreement over territorial conquest? Here, Teschke is less useful, primarily because of his rather exclusive focus on landed accumulation. While this helps to explore gaps in the story of the evolution of inter-unit politics in Europe, it is not as helpful in the theatre of European interactions with South Asia, where the decisive dynamic was mercantilist: capital accumulation based on trade and not land.

Having opened with a broad discussion of the nature of British sovereignty over a broad period of two centuries of Company rule, the remainder of this discussion will hone in on the very beginning of this period. In what follows, I argue that on the sub-continent, the process of establishing agreements about who could rule what territory, and how these territories could be ruled, was shaped by a common acceptance of a set of norms of mercantile governance. Put simply, Europeans and non-Europeans shared a belief that the legitimate ruler over a trading post was the actor who could guarantee that they would use such territory to pay rents on wealth produced through the carriage and exchange of goods to imperial overlords based in London, Agra, Golkonda and elsewhere. Crucially, although Europeans were commonly able to cooperate in various ways, their common cultural heritage rarely provided the basis for more rule-governed behaviour amongst them in relation to territorial expansion – with alliances and more temporary accommodations frequently disturbed by conflict and political manoeuvring. Furthermore, within this minimalist set of guidelines, power asymmetries were not necessarily expressed as a European consensus on what territory non-European powers legitimately owned, or how they should rule that territory, but rather which forms of rule, and in what areas, the imperial centre was responsible for the ruling prerogatives of its agents. Proceeding along these lines, Britain's increasing power in India was generally expressed through the piecemeal usurpation of indigenous rulers' sovereign prerogatives rather than the immediate imposition of European-style identities on non-European actors.

Going further, this chapter shows that the key points of transition in this order did not necessarily correspond to changes in the relationship between land

and power. Drawing upon a broadly Weberian perspective of the development of state power, this study aims to show that European power was not necessarily practically exercised by metropolitan sovereigns. Put simply, the British Crown, having alienated much of its responsibility for East Asian affairs to the Company, had little influence over decisions regarding when and where it was appropriate to expand its jurisdiction in India. Even the directors of the East India Company in London did little more than rubber-stamp the decisions that were being made on the ground. The changing balance of power between sovereigns and their own agents, as European states acquired the capacity to discipline their own colonizing agents, generated changes in the principles and patterns of expansion in India.

This shift in our analytical and historiographical assumptions means that the shift in balance of power between Britain and the Mughal Empire did not immediately change the norms of sub-continental international society that emerged between the Europeans (here, the British) and the indigenous rulers of India, which were based on a shared desire for financial and material gain. After Lord Clive triumphed at Plassey in 1757, territorial expansion continued along similar lines in India for another 100 years, because the British imperial state continued to possess limited control over its sub-continental agents. Change in the Indian international society only came later, when the most powerful state (Britain) took on a bureaucratic-developmental character. This process culminated in the 1857 Mutiny when the Empire threw off the cloak of Mughal sovereignty and institutionalized a policy of paramountcy in relations with other Indian states. Crucially, this was less motivated by a need to impose traditional *European* standards of civilization, and more compelled by the need to impose new *British* norms of governance.

Common principles of commerce

The Company Charter: expansion in European norms

Understanding the manner in which British agents negotiated the terms of territorial conquest with South Asian rulers requires some attention to the specific constitutional structures that had been established to regulate this process. The British project of discovery and commerce in the Indian Ocean was a very 'bottom-up' affair, having been initiated by a consortium of mostly London-based merchants as a profit-seeking venture. Following on from the example set by their Dutch counterparts and several other British projects of discovery and trade in the North Sea and the Levant, the merchants petitioned the Crown for permission to set up a chartered company that would establish and manage trade around the Cape of Good Hope, in particular, competing for control of the lucrative spice trade of the East Indies.

The East India Company was a semi-public body. Corporate strategy of the Company was managed by a Directorate in London that answered to the Company Board and a community of shareholders. The Crown awarded the Company monopoly rights to engage in trade and diplomacy between South Africa and Indonesia, subject to periodic renewal and revision of the terms of its Charter. Meanwhile, day-to-day affairs abroad were managed by a series of Company factors, who reported to London *via* a 'Presidency' town (usually the key trading post in that region).

To the contemporary observer, one of the most striking aspects of the half-century of Company trade after its founding charter in 1600 is the remarkable lack of concern for how lowly factors might absorb territory on behalf of the Crown. While the original Charter covered in great detail how decisions would be made at the Company headquarters and on Company ships, and specifically provided for the Company to acquire the necessary land to set up its premises in England, it did not specifically provide for the acquisition or cession of sovereign authority abroad. Part of the reason for this apparent gap stemmed from the fact that the British rulers who would ultimately become sovereign over those lands that the Company controlled, were relatively uninterested in *how* the Company acquired its share in the trade to the East. Throughout the seventeenth and eighteenth centuries, the main question for the Crown was how the Company could begin to generate profits for British fiscal reserves, which had been radically depleted by successive wars.

The other concern was whether revenue could be raised without upsetting relations with other European powers (and possibly instigating more conflict). For Parliament, an institution that was becoming increasingly assertive over the course of the century, the main issue was how individual parliamentarians could augment Company profits (at this stage still prospective) and secure Company property, particularly if they were shareholders or the clients of such, or how they could acquire a share in the trade if they were not. Crucially, few members of the London elite knew much about East India trade or about the sub-continent itself and preferred to intervene in decisions about institutions and finance located in the capital, rather than fiddle in the *minutiae* of overseas strategy.[3]

Throughout the seventeenth century, a succession of new charters[4] experimented with the Company's constitutional framework in order to make it more profitable and flirted with the idea of opening up the trade to other British trading companies and consortia. Once the Company had acquired a significant number of factories and then forts, its representatives could then use these acquisitions as bargaining chips in such debates, to justify the exclusion of 'interlopers' (who possessed no such property) from the Company monopoly. However, the fact that the Company had become a governing body on the Indian continent simply did not, in and of itself, register as a significant issue in public debate in London until long after these prerogatives had already been exercised de facto, and de jure authorization had been acquired from rulers abroad. The only significant restriction on the Company's generous trading monopoly was an explicit prohibition from encroaching on the trading settlements of Christian allies, which might precipitate conflict in Europe, and the more general rule, applied to all Chartered entities in Britain and abroad: that their laws not 'be repugnant' to English law. In this respect, 'European norms' barely applied.

'Old corruption': the underlying ethos of expansion

While territorial expansion did not feature as a major item of discussion in public debate in early seventeenth-century London, it *was* a source of dispute between the Company governors in London and their servants abroad. The directors were predisposed against territorial aggrandisement (Bryant 1998: 32; Berger 1990: 60–62). Indeed, one of the main reasons why territorial sovereignty may not have figured in the original Charter was that the directors did not immediately recognize the importance of such a trading strategy. Hence, the dictum of the first Ambassador to the Mughal Court, Sir Thomas Roe (cited in Gledhill 1964: 4): 'Let this be received as a rule that, if you will profit, seek it at sea and in quiet trade; for without controversy it is an error to affect garrisons and land wars in India.'

There were also some institutional factors that militated against the type of investment strategy required to make such an approach work. The directors were immediately responsible to their shareholders who, while having little understanding of overseas trade, consistently held management to account regarding the payment of dividends. The upshot of this argument is that while the Company was by no means philosophically opposed to military force, over its first 100 years, when profits were not always guaranteed, it preferred the use of naval supremacy and other forms of coercive threats over fixed investments in forts and landed warfare (Marshall 1998: 5–6).

The Company directors found it difficult to hold true to these maritime preferences. In part, this was due to the determining influence of distance, though distance understood in its capacity-limiting role, rather than as a socializing force. While the directors were interested in foreign policy, they were dependent upon their employees' judgement and interpretation of the strategic and mercantile status quo abroad and, due to the huge distances and slow pace of communication, could not exert direct control over the day-to-day management of Company affairs until faster ships and routes and (much later) wire communication were established. Factors could therefore often present their decisions to establish relations with local rulers, make and break treaties, and build fortifications as faits accomplis, and support these arguments through reference to a range of immediate exigencies.

Of greater importance was that the socio-economic *mores* of the day specifically prevented the directors from adopting a more hard-lined approach to the use of Company funds for unauthorized expansion. Corporate and governing practices in London at the time remained beholden to a neo-feudal system of 'old corruption': ' … a parasitic system that taxed the wealth of the nation and diverted it into the pockets of a narrow political clique whose only claim to privileged status was its proximity to sources of patronage' (Harling 1996: 1). The British did not lack a sense of morality in public affairs.

However, it was accepted that members of the elite were entitled to bend these principles in the service of family, friends and colleagues. Inter-personal linkages, on the one hand, amongst individual members of the Company Board, shareholders, creditors, English exporters and their respective families, and on the other, between these London alliances and the employ abroad, influenced the way in which the organization accumulated wealth (Sutherland 1952; Chaudhuri 1986: 105). The directors could neither lay down a stringent set of policy dictates regarding expansion, nor effectively censure all those who disobeyed because the individualistic exploitation of Company resources was the *raison d'être* of a Company job. Through the exercise of employees' limited rights to engage in personal trade, Company employment was seen as a potential way of acquiring a personal fortune. Many applicants drew heavily on personal connections to the Board of Directors to acquire their positions. These patrons, in turn, often had the same understanding of the relationship, and generally obliged clients' requests for jobs in the expectation that new employees would repay on return (once they had established their own fortunes).

These dynamics came into play when opportunities for landed expansion emerged.[5] For the average senior Company servant who was attempting to gain entry into the vast commercial networks of the sub-continent and to exploit niche markets, there were significant advantages to be gained by stretching and even contravening London dictates on this issue. While European powers' best route to trade monopoly was probably through the assertion of their growing naval supremacy, gaining a permanent place in domestic markets required, at the very least, a toehold grip on the Asian coastline. In more peaceful and predictable regions this could be achieved by the posting of a permanent 'factory', composed primarily of administrators and clerks, in key port towns.

However, these factories' capacity to generate a profitable trade was necessarily limited by their dependence on local and other European rulers for security, their reliance on 'go-betweens' to procure goods, and the British factors' lack of standing in local markets compared to other, more established European traders. Incremental improvements could be achieved in the terms of trade through subtle diplomacy and displays of naval prowess,[6] but these gains would always be uncertain, particularly given the Company's lack of knowledge of, and inability to predict changes in, inland political formations and alliances. For Company servants, who were interested in establishing personalized trading networks in order to enrich themselves and their patrons, and who owed few direct responsibilities to the shareholding community at large, this meant that there were strong incentives to try to improve the Company's position by seeking more permanent, fortified establishments.

'Corruption' in India: a perfect fit for an international order?

Local Indian rulers had good reasons to facilitate the establishment of such settlements, particularly in the more fragmented political systems of the south. There was significant competition over relationships with international traders in general, and European traders in particular. These vast networks created opportunities to access a range of military technologies, with guns and horses at the top of the list. For a courtly culture that placed a high premium on the symbolic display of wealth, the acquisition of foreign curiosities was also significant. Local rulers also had incentives to promote the growth of commerce generally, because taxation of the movement of goods was often one of their largest and most easily augmented revenue streams.

Crucially, the authority to decide whether Europeans were granted these privileges was generally deferred by kings and emperors to local fief-holders (*nayaks*). The idea that an allied sea power with minimal apparent capacity for expansion inland would fortify a heretofore unprotected part of the coastline at minimal cost to local coffers, tended to outweigh the *nayaks*' fears about permanently alienating territory. In addition to these factors, the perception that European traders would contribute to the broader development of military infrastructure appears to have outweighed the consideration that the traders were a 'foreign' force.

Most peculiar to modern nationalistic sensibilities, and contrary to the European approach at the time, was that unlike their British counterparts, Indian rulers appeared to have attached minimal importance to the prospects of their subjects becoming subject to an alien system of authority and law. One of the founding principles of most British Charters to overseas companies was the notion that British subjects would be able to avail themselves of British justice. In comparison, Indian rulers were willing to hand over considerable rights to enforce commercial and penal law within European enclaves. The upshot of this outline is that the *nayaks*, who were often directly engaged in deciding whether and how British expansion might begin in India, were open to a range of accommodations, as long as they could retain the favour of their ultimate overlords.

To summarize, this image of decision making about the expansion of British power on the sub-continent contrasts significantly with standard English School accounts. While Company servants had specific reasons to avoid conflict with other Christian powers, this only consisted of a proscription to prevent disagreements with allies of the Crown. A common European 'culture' tended to influence decisions, but more in terms of identifying the techniques that were necessary to develop successful trade rather than prescriptive guides for the judgement of the merits of prospective indigenous sovereigns (the 'standard of civilization' case being a typical example of this). While distance was a significant factor in these negotiations, it was less important as a source of (de)socialization to international norms and more a capacity limitation, exacerbating London's inability to control Company servants.

In this context, principles that were *shared* amongst Europeans and non-Europeans alike regarding the nature and purpose of trade constituted the guiding rationale of mercantile territorial expansion. Both European and non-European sovereigns adopted a largely hands-off approach towards the activities of the East India Company, viewing the institution as a largely instrumental means of augmenting wealth, security and standing within their respective spheres of influence. This does not mean that *purely* instrumental factors decided when and where expansion would begin. The 'old corrupt' nature of employment and social advancement in London society, and the ostentatious Indian approach to wealth creation decided how their shared interests in the rents of commercial transactions would be translated into concrete arrangements over territorial rule. This also does not mean that there were not significant cultural differences as regards the understanding of these transactions, as suggested by the different approaches to subjects and foreign laws.

Beginning territorial acquisition

Surprisingly, the question of when the British first acquired ruling prerogatives on the Indian sub-continent has not received much attention in the literature. For the most part, scholars have been willing to take the establishment of the factory at Surat in 1612 as the beginning of British settlement and/or to accept Lord Clive's larger acquisitions of Mughal fiefdoms around Calcutta in 1757 as key 'markers' of British expansion. While these bookends are useful, they fail to pinpoint the precise moment when Company officials began to exercise political authority over Indian territory and the people resident therein. For this, others have suggested that we look to the first establishment of a fortified settlement (Bowen 1988: 160; Watson 1980). In this type of establishment we see the Company providing protection and administering justice: hallmarks of territorial authority.

For constitutional historians, the founding of Fort St George at Madras in 1640 is the landmark. The *firman* (fief) granted by Damela Vintutedro Nayak, which included 'full power and authority to govern, and dispose of the government of Madraspatam', certainly looks like a transfer of political authority over territory.[7] The offer to 'surrender the Town called Madraspatam and all the Ground belongeth to it, the Government, and Justice, of the Town into your [the Company's] hands',[8] which came in the confirming *firman*, may be seen as signalling a qualitative shift in the Company's competencies from that of a pure mercantile organization to a mercantile company with genuine responsibilities to govern inland. If few people at the time, including the immediate participants, saw any great portent in the negotiations that concluded in the founding of Fort St George in Madras, the phase was crucial, because it saw the Company move from an exclusively trading organization based in factories that had been planted on the soil of other rulers, to a power possessing (relatively) exclusive ruling prerogatives vis-à-vis specific portions of land, including the authority to govern subjects of the indigenous rulers. At issue here is the question of who were the legitimate rulers and for what purposes they should exercise governing powers.

The task of tracing these origins does not end here, because although Madras is clearly the first *surviving* Company establishment of this sort on the sub-continent, it was not the first fortified settlement. In this respect it was the successor to Armagon, a failed attempt to establish a fortified town at a location to the north of Madras.[9] Crucially, the decision-making process that led to the de jure establishment of Fort St George was intimately connected to the failed de facto efforts to establish a similar settlement at Armagon. Hence, an adequate study of the origins of British territorial authority should not simply concern itself with Madras (1639–53), but should extend backwards to include some consideration of the reasons why Indian-based factors built Armagon and why they dismantled their settlement there (1625–39).

The failed settlement of Armagon

This story begins at Masulipatam on the Coromandel (East) Coast of India. The Masulipatam factors had originally received instructions from London to settle here on the expectation that they would procure the woven and patterned cloth that were specialty products of the region.[10] Indian textiles had relatively quickly emerged as a crucial item of trade for the Company. The Dutch, who dominated the spice trade, soon discovered that Indian textiles were in great demand across South-East Asia. By procuring textiles from India, they could acquire spices in South-East Asia to send back to Europe and reduce their need to trade precious metals brought from home. While the British were able to procure textiles in other towns like Surat, South-East Asian tastes for Indian textiles were highly sophisticated and diverse and the Dutch successes in procuring fashionable goods on the Coromandel Coast encouraged the British to follow suit (Sen 1962).

When the British first arrived at Masulipatam in 1611, they procured an agreement from the local *nayak* of the king of Golkonda to establish their factory and trade there. However, their successors consistently complained that the *nayak* acted in various ways to interfere with their trade: he had limited their freedom to trade by restricting them to do business at Masulipatam and appointing their intermediaries with merchants; he imposed excessive and additional taxes on their trade; and he had prevented them from settling their grievances with local merchants and producers.[11] So, for the first few years, Masulipatam was a failure for the Company, and the factors based there spent most of their time and energy on their private trade (Foster 1902: 1).

It is not entirely clear who initiated the discussion of whether the British should abandon Masulipatam. Some correspondence refers to letters that were sent from a *nayak* near Armagon entreating the British to settle there.[12] This suggests that the Masulipatam factors may have planted the idea in the minds of Bantam officials. Whatever the case, in 1625–26 the Bantam Presidency sent orders via *The Rose* for their merchants to begin searching to the south for a more suitable place to procure cloth. Both Bantam and Masulipatam factors looked to Dutch successes to justify the move south. Over the same period, the Dutch had also established a factory at Masulipatam, but experienced similar problems with the local *nayak*. They were better served by their fortified settlement in Pulicatt, which they had acquired from the Portuguese in 1609. Pulicatt was located much further to the south. Based on their problems with the local *nayak* at Masulipatam, the British factors had concluded that these fortifications had been a vital means by which the Dutch had cornered the cloth trade because it helped to ensure their autonomy and protection from local rulers on the coast.

The Rose was wrecked on the coast near Masulipatam in early 1626, but the merchants rescued the instructions and sent an emissary in the salvaged parts of the same ship to investigate the prospects of a settlement at Armagon. Some 11 days later the emissary arrived and made contact with a group of local merchants who welcomed them and encouraged them to settle there. The same merchants also assisted the Company in procuring a *cowl* (written grant) from the local *nayak*. This agreement acknowledged their rights to settle and gave them a deal on customs and the right to mint coins but, according to the servants' reports, included no word on issues of the Company's power vis-à-vis local merchants or rights to fortify.[13]

Even at this early stage, there were signs of trouble ahead. The factors had to pay an 'ordinance' – probably a 'gift' rather than a formal tax – to procure the *cowl* and the Dutch were being spoilsports. A 'Brahmin' ambassador of the Dutch had interfered in the negotiations, informing the *nayak* that the English would not be capable of building a sufficiently large and wealthy settlement, and therefore should not be entertained. The factors were at best ambivalent in their representation of the character of their European and non-European competitors. In the same communication that cursed the untrustworthy and greedy heathens, they bemoaned the lack of Christian unity shown by the Dutch, who 'like greedy gluttons, will rather stuff their stomachs 'till they surfeit, rather than spare (if in their power) the needy the least morsel'.[14] While Christian origins may have encouraged expectations of good behaviour, they certainly did not necessitate a favourable judgement about moral standing and commitment.

Despite the early warnings of trouble, the Company factors were fed-up with Masulipatam and encouraged by the signs of a safe and strategically located port and the good access to cloth, so they began the move to Armagon. It is not clear whether they initiated this move with the intention to build a fortress there. The preceding communication and permission from Bantam had made no reference to such plans. However, given that a communication to Bantam, written a little more than two months after landing, noted that the brick and lime walls of Armagon were already finished, it appears as though the settlers had always intended to fortify.[15] This action initiated a long campaign by the Coromandel factors to encourage Bantam to provide monetary and material support to add to the fortification. Now was not the time to be sparing, especially when faced by threats and betrayal by both the land powers and the Dutch. With regards to the former, the servants were mindful of their recent departure from Masulipatam. This 'escape' was undertaken without the *nayak*'s knowledge or approval, and resulted in a heated dispute over outstanding accounts. The affair seems to have been concluded only when the Company, acting with sanction from Bantam, stole a good portion of the Masulipatam fleet and held the *nayak* to ransom.

The security situation at Armagon provided the Company and the Coromandel factors with little comfort, however. A domestic war threatened the lands of their new benefactor there and the Dutch continued to harry their ships and undermine their fortunes in the *nayak*'s court. Here, it is interesting that the Coromandel factors gradually developed a more complex reading of the *nayak* over the course of their first year in Armagon. Rather than a characteristically duplicitous native, they came to view him as a man capable of rising above these origins: '[he] stands more upon his promise, than any covet of gain; which is a rare quality to be found in a Heathen.'[16] The problem was less the *nayak*, than the fact that he was surrounded by a court of intrigue, where both the Dutch and his own courtiers encouraged him to undermine his good faith. Perceptions of interlocutors' intentions were as likely to be based upon experience in negotiations as presumptions about religious/cultural affiliation.

As the Coromandel factors began to contemplate their prospective military power in Armagon, their requests to add to its fortification reveal a rapid assumption of responsibility to provide protection for the local population. While the need for the security of Company goods, possessions and personnel was initially paramount in their decisions to fortify, they began to envisage that the Company could and should take care of the security of its suppliers and clients. At the same time as they made their appeals for cannon, materials, engineers and money to assist in the completion of fortifications that would help to ward off landed armies and thereby help them to stay competitive with the Dutch, they also emphasized the importance of shielding the townspeople from marauding armies and attracting more local inhabitants to settle in their vicinity through promises of a more secure residence. While the Coromandel factors seemed blissfully ignorant of the complex politics that this would leave behind for future generations of Company servants, they were quite aware that in asking for money to fortify, they anticipated becoming a ruling authority of sorts. One communication expressly motivated for additional resources to fortify on the grounds that, 'in time, we should get the whole government of the place into our own hands'.[17]

Whatever their perception of the merits of this strategy and its long-term impacts, senior Company officials appeared unconvinced of the prospects of the coastal settlement plan and issued orders soon after the settlement of Armagon (it is not clear whether this came from Bantam or London) to dismantle all the Coromandel factories. Then, somewhat surprisingly, the Coromandel factors directly disobeyed this order. Instead, they consolidated themselves temporarily at Armagon and the Dutch town of Pulicatt until the effects of their piracy at Masulipatam were felt, and they could procure a new *firman* for a factory there. Ultimately, Armagon proved to be a poor settlement, plagued by the same trading problems as Masulipatam. Furthermore, since the factors were constrained by the apparent unwillingness of their directors to invest, the defensive infrastructure remained poor. By the latter part of the 1630s, when the issue had moved to the consideration of the prospects of settlement at Madras, everyone involved had accepted Armagon as moribund, with Coromandel's reference to a 'ruinous building'[18] matched by the new Surat Presidency's description of the 'mock fort'[19] located there.

Of greater significance is the degree of leeway that the Company servants had exerted in initiating this acquisition. While consistently constrained, and sometimes provoked by their superiors, the Coromandel factors had independently motivated for, and ultimately established, a fortified settlement on the coast. In doing so, they had not acquired major powers from the local ruler other than that of minting coins, but had quickly moved in terms of their self-identity from the position of mere traders to providers of protection. Ultimately, their ambitions to establish a fully functioning fort were not realized. However, this process set the stage for the negotiations over an acquisition that would lead London by the nose into a position of lasting and de jure political authority in India.

Founding Fort St George

Given the apparent failures of their previous two settlements, it is with some surprise that a few years after moving back to Masulipatam (leaving a skeleton staff in Armagon), the Company found itself embroiled in a project to establish a third settlement on the Coromandel Coast. Unlike the silent failure of Armagon, the establishment of the fortifications that would come to be named Fort St George became the subject of protracted censure proceedings in London and a blame game amongst overseas servants.[20] Hence, the individual characters in this story acquire greater significance than their counterparts involved in the Armagon acquisition.

Again, the story begins with a *nayak* to the south of Masulipatam, luring the factors to visit and investigate the prospects of settlement with promises of access to good cloth. The key player in this saga, Francis Day, received this offer from an emissary of the *nayak*, Damarla Venkatadra (or Venkatappa). Day sought permission from Thomas Ivy, then head of the Masulipatam factory, to head south and investigate. Despite having only recently (and quietly) deferred orders (again, it is not clear if these were from Bantam or London) to dismantle Armagon, Ivy conceded to this request. In 1639 Day made his visit to the south and returned with surprisingly (and as we will see, also suspiciously) good news about a small settlement called Madraspatam. In a letter that he personally delivered to Thomas Cogan, who had by then replaced Ivy as head of the factory, he all but begged for the Company to take up this offer. He reported of an abundance of quality cloth for sale, and a *firman* empowering the Company to set up a fort and settlement with, crucially, an agreement to front the capital for the fort. In addition, they had been offered good terms of trade, tax concessions, the *nayak*'s assistance in settling accounts with locals, rights to mint and, importantly, 'full power and authority to govern, and dispose of the government of Madraspatam'. In a classic 'motivation' letter, Day stacked up all of the arguments in favour of settlement, including the prospect that the Dutch might beat them to the punch, and perhaps more tellingly, that he had no personal interest in the acquisition, with only the Company's interest at heart.[21]

After receiving the request, Cogan asked Day to head back to Madras with a horse as a gift to keep the offer open. Day, who was preparing to leave the Company employ at this time, bargained for Bantam to pay off some of his local debts in return for his agreement to accept the mission. Cogan then began negotiations to acquire permission for the settlement. Interestingly, although Cogan's request was addressed to the 'Company in England', in the first instance it only went as far as the Presidency, which had recently been relocated from Bantam to Surat on the northern part of the West Coast of India. Cogan employed some of Day's arguments in this request and then added two of his own impressions. First, he provided a list of reasons why, in his assessment, the local *nayak* had made such a generous offer: a) encouraging commerce will enhance his wealth; b) the fort will help guarantee his own security; and c) he will be able to acquire new goods for his military and his court. Second, he explained why, in terms of broader trading policy, the Company should acquire a genuine fortified settlement: a) they needed trade in some other goods so as to limit their sterling outlays; b) neither Armagon nor Masulipatam were profitable trading centres; c) the Dutch had used their fortified Pulicatt to dominate the trade; and d) the fortifications at Armagon were worthless. In summary, Cogan argued that it was high time that the Company got serious and realized that a long-term trading policy in the East Indies required a fortification of this sort:

> Why shall not we, in some things, imitate our inveterate and most malicious Enemies [the Dutch]? Which is; as soon as any goods is bought, in their private factories, send for it, in their small Vessels, to their fort at Pulicatt; by which means they keep these People so much the more in subjection, and still command their own.[22]

The Presidency of Surat, after apparently deciding not to wait for London's approval, responded to Cogan cautiously as regards the potential expenses involved. However, he casually encouraged the establishment of a new and fortified settlement:

> You seem to be fairly far on with things anyway, it all seems reasonable (so long as it is true), you seem to have thought things through … but don't liquidate Armagon or build St. George until you are well assured of the Naik's resolution to receive you, and assist in the fortification … If you go forward with it, do what you resolve on to purpose; and build no such mock forts as that of Armagon.[23]

In 1640 Cogan and Day travelled with this endorsement in hand to dismantle Armagon and establish the fort at Madraspatam. When they arrived, they found out that the *nayak* had no intentions of paying for the fort.

Suddenly, the primary rationale for settling at Madras had unravelled. Recognizing that they would need additional financial support to complete the work, Day and Cogan sent a request to Surat, who forwarded the request to London. At the same time, they decided to go ahead with construction, investing huge amounts of Company resources in the work. Understanding Cogan and Day's decision making at this point is difficult, and is clouded by the 'spin' generated as each of the actors involved in the chain of command sought to exonerate themselves.

A blame game began. The *nayak*'s position was that the interpreters had misconstrued his offer. Day's position was that the *nayak* was lying. Cogan's position was that Day had at best been foolish in accepting the *nayak*'s word and at worst had simply misrepresented the *nayak*'s commitment for his own gain. The position of the Company Presidency was similar, but they also attributed some of the blame to Cogan who, after all, had sought to convince them to support Day. All of these representations were communicated in one form or another to the directors in London.[24]

What is striking – at least for the contemporary observer – is that during this whole dispute, none of the sub-continental factors suggested that despite the clear failure of their previous venture at Armagon, the fort and settlement (with all their attendant responsibilities and implications for future Company activities) ought not to have been established and/or new terms reached and/or another site selected. Rather, the debate centred on the issue of who should be censured for the (seemingly unavoidable) overspend on Company accounts.

Unsurprisingly, the response from London was unsympathetic, condemning the Fort St George project (and possibly ordering it dismantled). Yet, when the Company Presidency in Surat received these instructions they did not pass on the message to the Coromandel Coast. Instead, they asked London to keep the fort, based on the same line of security reasoning that had been offered since the founding of Armagon: that the Dutch were similarly secured and that the fort would allow them to trade independently of local influence (Foster 1902). At some point, word of London's displeasure reached the factors at Coromandel, as did London's demand that the local *nayak* make good on his promise of payment. By the time they received this message, Day had left the Company employ. Cogan and his colleagues then scripted a reply pleading for an alternative solution, remarking simultaneously on how faithful the local *nayak* had been and contrasting this with how unpredictable *nayak*s and locals are in general. In addition, they directly challenged the implicit logic of the initial acquisition: that the *nayak* would pay for the fort. Instead, they argued that it made no sense to accept the *nayak*'s part-ownership of the fort:

> If these people build us a Fort, and pay the Garrison; in what security is your estate, and our lives? – surely, in none at all: for tis far more freedom to live without a Fort than within; unless the Fort be at's [sic] own devotion.[25]

London was not impressed. In 1642, with no great change to the status quo in Coromandel, the decision was made to censure Cogan over the affair. Cogan was indignant in his communication, to the point of complaining, along with his colleagues, over London's dithering: 'make a decision and let us be done with it!'[26] He organized his defence by preparing a letter in council with factors in Bantam that defended the Madras project as financially sound (but not necessarily legally established) and returned to London to face the music. The committee that assembled to investigate Cogan buried the whole affair. Instead of holding Cogan to account and demanding that Fort St George be dismantled, they voiced their disapproval of the construction but decided to keep the fort:

> although it were a very indiscreet Action, to go about the building of such a Fort, when the Company's Stock was so small; yet, if ever the Company have a plentiful Stock, it may be very commodious and advantageous for them.[27]

They then joined with Cogan in blaming Day for the entire affair. Day, who could no longer be formally censured for his activities, was an obvious fall guy. A year later he was reappointed (albeit for a short stint), without having faced censure, in a command position at the fort at Madras. All this time, the building of Fort St George continued, playing a crucial role in securing the British position on the Coast over the coming years. The Coromandel factors generally remained quite uppity about the whole issue, complaining to London of their 'unthankful employment' in the founding of Fort St George (Foster 1902).

While a forensic examination of the motivations and justifications of the building of Fort St George is impossible, on the available evidence we can make some fairly rough estimates of how this acquisition took place. Day was struggling at Armagon and saw no prospects of improving his personal trade there. After scouting around for an alternative site, he made contact with the *nayak* at Madras and began a campaign to establish a settlement where he could make some decent money before returning home. In doing so, he probably did not blatantly lie to his superiors. The *nayak*, who had every reason to believe that attracting the English and establishing good relations was more important than the terms of the agreement, may have intimated commitments he was not prepared to fulfil. Clearly, in this respect, he was successful, though it is not clear whether he rested on the gullibility of the factors or knowledge that they could adeptly side-step London's sanction. Day then gave a favourable interpretation to the *nayak*'s promises. Day knew how to set the administrative wheels in motion, both with Ivy and then Cogan, who were also based at a moribund trading settlement. Both were astute observers of the successes of their Dutch and Portuguese competitors and were keen to buttress the Company's position. As men who were higher up in the Company and set on more promising careers, it is likely that they had Company interests in mind. However, they were also capable of manipulating communication with their superiors, by raising the spectre of European competition and local threat. As regional experts, and traders in their own right, they felt completely assured of their decisions and resentful of the interference of London bureaucrats who knew next to nothing about India.

Similar dynamics influenced the Surat presidents, but their main aim was to ensure that expansion continued while keeping their hands clean of the dirty work on the coast. This explains their judicious management of information flow between Coromandel and London. Meanwhile, the London administrators, who ultimately had to explain this acquisition to the shareholders, were deeply concerned with the expenditure but less in the sense of having to discipline those responsible for the outlay and thereby setting a precedent for future attempts of this sort. They were more concerned that their shareholders would see them as being active in reigning-in reckless expenditure in the East.

The mock trial of Cogan is the denouement for this buck-passing fiasco. Importantly, all of these actors recognized that ultimately the building of the fort in Madras was in their collective interests and that of the Company, and none of them questioned whether they had a right to take this possession, to transform the Company into a landed military force, or begin to acquire a set of sovereign prerogatives. Almost certainly, none were concerned with the question of whether the Indians were fit to govern themselves, as this would have been anathema to the very nature of exchange of prerogatives implicit in the negotiations. Finally, the Crown did not so much as raise an eyebrow as the Company took the first of what would turn out to be many little bites out of the sub-continent over the next century. The Coromandel acquisitions outlined in this section were never authorized by English sovereigns, but were subsequently sanctioned in Cromwell's Charter of 1661.[28]

Analysis and summary

Although limited to a small snapshot of early British settlement in India, the preceding account has significant implications for our understanding of the constitution of international society in the pluralist period and beyond. While this account does not cast doubt on the notion that many European powers may have shared a set of cultural understandings about the nature and purpose of political authority, territoriality and their attendant prerogatives, it shows that these understandings tell us relatively little about British expansion in India in the seventeenth century, a process not built on Manichean continental dichotomies but negotiated on a shared understanding that European and non-European jurisdictions could substantially overlap. What mattered was that the agents of sovereign powers on both sides could more or less invent the templates for mercantile outposts, attaching significant ruling powers to these entities as they went along. They were given this latitude in part because their superiors were directly interested in profits.

However, the fit between the interests of the British Crown and Indian counterparts is not the whole story. More significant is the fact that small alliances of well-placed individuals could use constitutional frameworks such as the Company Charter and the imperial *firman* to further their own wealth-generating agendas. The 'international society' that determined the course of early British conquest in India was a transnational network of individuals, not a European society of states. Yet, it demonstrated that actors from different cultural backgrounds were more than capable of negotiating common frameworks for interaction in the absence of outright European imposition – even though this may have been primarily motivated by desire for personal gain. Throughout history, the regulation of economic exchanges (be it in the form of 'tribute' in East Asia, or for the purposes of subsistence) have proven to be powerful motivational forces for the establishment of social relations between different polities (Buzan 2004; also see Schouenborg 2013), and this indeed seems to be the case between the British and their Indian counterparts, even though the 'depth' of these 'shared norms' during this time could be open to debate.

This explains why when Cogan returned to London, he did so in a spirit of indignation. He had been a faithful servant and helped to enlarge the Company coffers, so what was all the fuss about Fort St George? Of course, in later years, returning servants could not adopt such haughtiness. Take, for example, Warren Hastings, who was subjected to a trial after returning from India as its first governor-general in 1792 – mostly on the grounds that he had expanded British power in an inappropriate way. The more vociferous attack on Hastings is explained by the fact that a new understanding of British sovereignty had taken hold in the capital. However, again, this had little to do with broader principles established as part of a 'European international society', or even a British or English cultural norm. As Cain and Hopkins's (1980, 1986, 1987, 2001) analyses point out, the main changes taking place in how imperial acquisitions were evaluated and governed had to do with a cultural change taking place in *London*, resulting from the simultaneous rise of a financial and services class and the alliance established between these groups in the City. The definitive principle of public power to which this coalition held was of financial probity (not sovereignty or any other English School trope), and it is for this reason that reckless expansionists like Hastings could now be viewed negatively as 'corrupt'.

We do not have the space to explain this more rigorous approach to conquest here, but suffice it to say that it involved the subjugation of the Company to Parliament, the professionalization of the Indian bureaucracy and the insertion into government of the idea that the Crown's responsibilities on the sub-continent should be exercised with a 'public good' in mind, which, albeit partially, included the interests both of the English nation and that of her subjects abroad.

By widening our scope even further, we can begin to address the more contemporary aims of this collection to understand the implications of a hypothesized return to a pluralist international society. To be more specific, this study can help us to refine the types of question we ask and where we look for answers. This study suggests that we need to consider norms covering geo-political expansion as a fundamental question of any international society rather than, as has been commonly presented in the English School literature, a process that only takes place after questions of whether non-European actors are legitimate territorial sovereigns have been resolved. Clearly, the question of whether there was a legitimate sovereign in India mattered little to European decisions about expansion, and these events can only be seen in this light if we impose our own contemporary understanding of international relations on the past.

So, for India and China, the question may be less about what sort of standards of international society they may produce and whether their understanding of political expansion will differ substantially from the anti-colonial consensus that has bound the expansion of European, but more importantly North American power since 1945. Answering this question then requires a different approach to that suggested by the culturally (or at best, geographically) defined concept of international society that dominates English School thought. According to the account above, culture matters in deciding how expansion proceeds, but less in the sense of the underlying civilizational principles determining who is allowed to be a sovereign and who gets conquered, and more in the sense of: a) a deeper social consensus about the nature and purpose of the public institutions that exercise political authority; and b) a practical assessment of whether central actors are in charge of the processes of expansion that occur within their broad jurisdictional ambit.

From this perspective, our forecasts of what a future pluralist international society might look like should not look to what we may perceive as the 'classical models' of international relations in China and India, in the hope that these will provide us with some essential secrets as to how these powers will act in the present. Instead, we should be looking to how these contemporary 'societies' understand the role of the state and particularly its responsibilities to motivate for their interests abroad. By looking for key differences between each society, and between both societies and international standards, we could gain a sense of how attitudes and approaches towards international expansion, and particularly neo-colonialism, might change in the future.

Notes

1 For some examples of works that attempt to capture the indigenous perspective see: Sudipta Sen, *Empire of Free Trade: The East India Company and Making of the Colonial Marketplace* (Philadelphia, PA: University of Pennsylvania Press, 1998); Lakshmi Subramanian, 'Arms and the Merchant: The Making of the Bania Raj in Late Eighteenth-Century India', *South Asia-Journal of South Asian Studies* 24(2) (2001): 1–27; 'Banias and the British – the Role of Indigenous Credit in the Process of Imperial Expansion in Western India in the 2nd-Half of the 18th-Century', *Modern Asian Studies* 21 (1987): 473–510.

2 Between 1600 and 1650 factories were established in places as far apart as Basra (Iraq) in the West and Hirado (Japan) in the East. See the map in Lawson (1993: 180).
3 For the argument that the shareholders showed blind faith in the value of their stock as a long-term investment that paid reasonable dividends see Marshall (1968: 27). Laxity with regard to India may be contrasted with the shareholders' regular contestation of issues pertaining to the management of East India House in London (Bowen 1989: 192; Chaudhuri 1986: 100). Elections were on a yearly basis and frequently subject to manipulation on behalf of the principal shareowners. For the argument that the Board's authority was further consolidated in the hands of the chairman, see Marshall (1968: 26–27). For the argument that most parliamentarians knew little as well, see Bowen (1988: 160).
4 These charters were regularly renewed, with some important exceptions, at 20-year intervals.
5 This struggle between individual and corporate interest came into play in many decisions about the use of Company resources. One example was the question of what portion of homeward cargo would be devoted to personal *vs.* company goods.
6 A good example here is the victory over Portuguese forces at sea on the west coast, which played some role in acquiring the Mughal permission to establish a factory at Surat.
7 *OC – Factory Records Miscellaneous 9, Document A 'Firman granted by Damela Vintutedro Naiqe to Francis Day' – Chief of Armagon factory on behalf of the East India Company'*. Importantly, this acquisition did not signal a quantitative shift in the Company's activities from an organization primarily concerned with trade, to one that was mostly occupied by governance issues (taxation, security, justice, etc.), a tag that is still rightly attached to Clive's acquisitions.
8 *OC – Factory Records Miscellaneous 9, Document F 'Firman to Thomas Ivy'*.
9 This clarification, which has been commonly neglected in the literature, has been stressed by an anonymous transcriber of the OC from the founding of Fort St George: 'Armagon, and not Fort St. George – must be considered the first fortress possessed by the British on the Coromandel Coast', *OC – Factory Records Miscellaneous 9, 'Anonymous prefatory Remarks'* (and to that we may add, the first British fortress in India). Much of what follows is deeply indebted to William Foster's account and his efforts to preserve and catalogue the OC (1902).
10 This instruction would almost certainly have been based on intelligence received from Bantam or Surat.
11 *OC – Factory Records, Masulipatam, 'Letter to the King of Golconda: signed George Brewen, Thomas Johnson and Lawrence Henley'* [attached to a consultation of 16 September 1628]. In their communication of these grievances to the *nayak*, the traders opined their lack of status in the region vis-à-vis other more established powers, questioning the *nayak* provocatively whether he thought only one nation (the Portuguese) could control the seas of the coast.
12 *OC – Factory Records, Masulipatam, 'Letter from Armagon to Batavia: Signed Thomas Johnson and Beaverly'*, 19 April 1626.
13 No original document has survived.
14 *OC – Factory Records, Masulipatam, 'Letter from Armagon to Batavia: Signed Thomas Johnson and Thomas Mille'*, 26 January 1626, p.48.
15 *OC – Factory Records, Masulipatam, 'Letter from Armagon to Batavia: Signed Thomas Johnson and Beaverly'*, 19 April 1626.
16 *OC – Factory Records, Masulipatam, 'Letter from Masulipatam to Batavia: Signed Thomas Mille'*, 3 June 1626 [transcriber's notes say this letter was actually written by Thomas Johnson].
17 Ibid.

18 *OC – Factory Records Miscellaneous 9, 'Letter to East India Company in England: Signed W. Fremlen, Fra. Breton, Ben Robinson, Jn. Wylde'*, 29 December 1640.
19 Cited in E. *OC – Factory Records Miscellaneous 9, 'Letter to Bantam: Signed Thomas Cogan, Henry Greenhill and John Browne'*, 20 September 1642.
20 It is perhaps for this reason, and the lasting significance of the Fort itself, that documentation of this tale of conquest has been more faithfully preserved.
21 *OC – Factory Records Miscellaneous 9, Document A 'Firman granted by Damela Vintutedro Naiqe to Francis Day – Chief of Armagon factory on behalf of the East India Company'*.
22 *OC – Factory Records Miscellaneous 9, Document C, 'Letter from Masulipatam to Company in England'*.
23 Cited in E. *OC – Factory Records Miscellaneous 9, Document E 'Letter to Bantam: Signed Thomas Cogan, Henry Greenhill and John Browne'*, 20 September 1642.
24 *OC – Factory Records Miscellaneous 9, 'Letter to East India Company in England: Signed W. Fremlen, Fra. Breton, Ben Robinson, Jn. Wylde'*, 29 December 1640.
25 *OC – Factory Records Miscellaneous 9, Document E 'Letter to Bantam: Signed Thomas Cogan, Henry Greenhill and John Browne'*, 20 September 1642.
26 Ibid.
27 *OC – Factory Records Miscellaneous 9, Document K. 'Report of a Committee appointed to hear Cogan's case: Wm Methwold Deputy; Ashwell; Mr Massingberd Treasurer; Mr Holloway; Mr Wilson'*, 13 May 1645.
28 *The East-India Companies Charter, Granted by the Kings Most Excellent Majesty Charles the Second*, under the Great Seal of England, 3 April 1661.

6 International relations in the Americas during the long eighteenth century, 1663–1820

Charles Jones

Introduction

Early sixteenth-century Spanish conquests were so spectacular and created such an extensive new empire for the Habsburg dynasty that it is easy to forget that most of the western hemisphere continental land mass remained beyond the reach of Europeans until the eighteenth century. Erase the coastlines, and the map of European settlement in the Americas looks more like Indonesia or the Philippines than a solid continental territory. Meanwhile resistance continued along several frontiers into the nineteenth and even the twentieth centuries, punctuated by periods of truce regulated by treaties that implicitly or explicitly recognized the continuing sovereignty of indigenous polities.

Also easily forgotten by those unfamiliar with the history of modern states in the Americas are the tensions that developed from the start between imperial authorities based in Europe and growing creole colonial populations, and which led finally to independence.[1] Yet limits can be set, both in space and time, to the theatre in which some possibility still remained that First Nations might secure enduring political survival in the face of inchoate creole states struggling to break out of their imperial cocoons. Accordingly the first section of this chapter is devoted to identifying some exemplary cases and establishing a tipping point in the late seventeenth century after which the outcome was hardly in question. This bracketing exercise forms a prologue to narratives of the long eighteenth century which occupy the three remaining sections, each of which examines one set of conventions for the coexistence of disparate polities across uneasy and ultimately unsustainable American frontiers, starting with the far south of what is today Chile, moving next to the south-west of Brazil, and concluding with the northern plains and forests of trans-Appalachian North America.

The first generation of the English School scholars worked with a clearly diffusionist model of modernity in which international relations were governed by a single set of norms and institutions that could be seen to develop over time within a steadily expanding European world (Bull 1995; Bull and Watson 1984a). The principal normative function of this approach was to help legitimize the transfer of power from Britain and other European powers to their African,

Caribbean and Asian dependencies that had recently taken place. Achieving such a clear vision was not without cost. It is mildly ironic that Herbert Butterfield, many years later convenor of the British Committee on the Theory of International Relations and ex officio doyen of the English School, should have established his reputation with a critique of what he called Whig history, the kind of narrative designed to justify the present state of things (Butterfield 1931). For the English School was established on the basis of a thoroughly whiggish account which marginalized institutions (such as political exile and dynasticism) that were bound up with, yet analytically distinguishable from, diplomacy and the balance of power in early modern Europe. It also required the exclusion from international society of polities outside the emerging European system that did not adopt European practices. Turkey or Japan might join international society, but their interactions with existing members were not regarded as properly social until there had been sufficient symbolic acts of cultural submission and practical mimicry.

An alternative view of international society, informing this volume, is grounded in the theses of varieties of modernity or multiple modernities (Schmidt 2006; see also Eisenstadt 2002; Bayly 2004). This cluster of interpretations accepts that polities beyond the European state system found their own paths into modernity, often unwittingly aided by the spread of European technologies and practices along trade routes, but sometimes by much more deliberate observation and emulation. The premise of this chapter is that relations between such polities and expanding settler states in the Americas constituted an international society worthy of the name even when these processes of modernization and normalization were still inchoate. Up to a point it was functional for modern states and their publicists to regard the Westphalian system as more than an ideal type (Krasner 1995–96; Osiander 2001) and as a unique form of political modernity, since insistence helped make it so.

However, the tipping point at which such aspirations became delusory was passed some time ago. To adopt the wider definition of international society advanced here has positive value in a contemporary world where modern states have constantly to deal with non-state coercive organizations that cannot easily be brought within their established institutions and do not share their values. The question posed by experience along the frontiers of creole expansion in the eighteenth century was how international society functions when the norms and institutions that guide relations between polities are merely convergent, and far from agreed. It is once again a pressing question in the twenty-first century.

Establishing the frontiers

Surviving indigenous polities within easy reach of the Spanish did not last long. Cosijopi, the Zapotec ruler of Tehuantapec, who had formally submitted to the Spanish on learning of the fall of Tenochtitlan, retained effective government of his realm by becoming a vassal of Charles V. However, following his death in 1563 the area came under crown control (Zeitlin 1989: 34). Also in Mesoamerica, the Mixtec states that had resisted Aztec overlordship up to the arrival of Cortés soon regretted their submission to the Spanish, but their rebellions, in 1528 and 1548, were defeated (Hamnett 1999: 72). By contrast, in those parts of New Spain that were relatively remote from creole settlement, indigenous states persisted 'long into the postconquest period, with their territories and many of their internal mechanisms essentially intact' (Lockhart 1992, citing Gibson 1962). However, the emphasis here is firmly on 'internal'. Any attempt to throw their weight around would almost surely have been promptly suppressed.

In the Andes, Manco Inca had been installed by Pizarro following the defeat of Atahualpa. As soon as he realized that he was not in fact sovereign, he set about organizing military resistance, coming close to success in 1536–37. Forced to lift their siege of Cuzco, Manco's forces withdrew and established an independent state some 200 km distant, around Vilcabamba, from which they continued to harass the Spanish for a generation, until their final defeat by Spanish forces in 1572 (D'Altroy 2002: 319–20). Further south, peoples of the Araucanian language group who had resisted Inca hegemony were conquered for a time by Spaniards hungry for labour with which to exploit nearby placer mines. Sustained resistance left virtually no Spanish presence south of the Bío-Bío river by the second quarter of the seventeenth century and obliged the Spanish to concede the independence of indigenous polities by a treaty of 1641, confirmed in 1655.

With these exceptions, the story throughout much of the hemisphere was one of steady encroachment by European and creole slave-raiders, missionaries, ranchers, officials and settlers into a political space already radically transformed by the indirect effects of European contacts. The Zapotecs, the Mixtecs, Manco Inca's followers and the Araucanians all represent lingering survivals of older political orders. The indigenous polities with which Creoles would struggle throughout the eighteenth century and into the nineteenth, including the Plains Indians of North America, the emergent Mapuche state in the far south-west of the continent, and many of the seemingly untouched Brazilian tribes may initially give this impression, but were in a fundamental sense *modern* polities: the outcome of direct or indirect contact with European culture rather than stalwart resistance to it. Indeed, Boccara (1999) resists any use of the term 'Mapuche' before 1760 and describes a process of ethnogenesis in which the Mapuche emerge only gradually from military cooperation between a congeries of culturally related, semi-nomadic bands over a period of more than two centuries.

So 1572 in the Spanish empire, and notional dates roughly a century later in the North American and Brazilian interiors, may be taken as the points at which a modern international society in the Americas, embracing both indigenous and creole polities, can plausibly be said to have formed. While embracing the whole land mass, this society, with its distinctive forms of interaction, was shaped and energized by advancing frontiers as Creole societies, their populations increasingly reinforced by migration and miscegenation, began to spill out from coastal enclaves and attempt permanent settlement in the continental interior.

By the second half of the seventeenth century, more than 150 years after the initial Spanish incursions, European dominion in the western hemisphere was still very far from complete, but the frontiers across which creoles and Europeans confronted indigenous peoples had been clearly established (Barraclough 1999: 179, 181; Hemming 1978: xii; Lombardi and Lombardi 1983: 28; Salisbury 1996: 417). French settlement had been established in the St Lawrence valley, extending inland as far as Fort Niagara by 1679. There was also a tentative French presence in the middle reaches of the Mississippi and in the Ohio valley, as well as on the shores of the Gulf of Mexico, where their eastward expansion toward Florida had been checked, in 1696, by the Spanish establishment of Pensacola, a military settlement designed to guard the flank of their tenuous settlements in northern Florida. British settlements stretched along the eastern seaboard in a not yet continuous strip from present-day South Carolina to Newfoundland, while in the far north trading posts had been established by the British Hudson's Bay Company at Rupert's House (1668) and Fort Albany (1679). British, French and Spanish traders and exploratory expeditions had ranged further, but by far the greater part of the North American land mass remained beyond European or creole rule, and such permanent settlement as there was may best be regarded as a landlocked archipelago. Though the indigenous peoples of the Great Lakes region had already been profoundly affected by the nearby European presence, and hostilities between French and British imperial forces were intermittent in this corner of the New World from as early as 1620, it would be only in the eighteenth century that Anglophone settlers crossed the Appalachians in any great number. Up to the 1680s it is reasonable to say that international relations in North America were 'dominated by the interactions of Amerindians with one another and with the invading Europeans, rather than by European imperial or inter-colonial rivalries' (Steele 1994: 23).

The position in Meso-America and further south was very different. Here the Spanish had been able to form extensive successor states following their conquests of the Aztecs and the Inca. Yet relatively little expansion of the territories of the indigenous empires had been accomplished. The northernmost limits of Spanish authority in the mid-seventeenth century still fell short of the Rio Grande in the west, though a finger of effective control reached northward, along the river's upper valley. To the south, Spanish control was largely limited to the uplands or *altiplano* everywhere north of present-day Bolivia, though the Paraguay-Paraná river system, down to the strategic outpost of Buenos Aires on the Atlantic coast, together with parts of the plains to the west and north of Buenos Aires, were lightly settled, while to the west of the Andes, the Bío-Bío provided a frontier, policed from the marcher city of Concepción, founded in 1550.

The Portuguese, for their part, had settled several enclaves along the Atlantic coast but had yet to penetrate the interior in any substantial or permanent manner. *Bandeirantes* had reached out into the southern interior from São Paulo from as early as the sixteenth century, hoping to repeat the spectacular 1545 Spanish discovery of silver mines at Potosí. Failing in this objective they made a business of enslaving indigenous people for sale to plantation owners on the coast, including those settled at the growing number of Jesuit missions in the interior. The *bandeirantes* reached what would soon become the captaincy of Minas Gerais, to the north of São Paulo, in the 1670s, but the first rumours of gold, which would lead to rapid settlement in the eighteenth century, only reached the coast in 1693. The very success of mining in the interior soon afterwards led the Portuguese Crown to prohibit further exploration, in the hope of preventing a drain of illegally mined gold down the Paraná valley with consequent loss of revenues. However, this official policy was unsustainable, and by the mid-eighteenth century *bandeiras* had reached deep into the central highlands, to Goiás and Mato Grosso, close to the headwaters of the southern tributaries of the Amazon river system.

Between Spaniard and Portuguese, a rash of autonomous Jesuit settlements ran northwards from the upper Paraguay and Paraná valleys during the first third of the sixteenth century, later spreading into Mato Grosso and western Amazonia, and as far north as the interior of New Granada in present-day Venezuela (Hemming 1978: maps 3–4). The eighteenth-century expansion of both Spanish and Portuguese settlement spelt the end for these extensive polities, which were attacked by Spanish and Portuguese forces from as early as 1756, culminating in the expulsion of the Jesuits from Portugal, France and the Spanish Empire in 1767, and the suppression of the order by Pope Clement XIV in 1773.[2] Whatever may have been the European causes of this development, its effect in the western hemisphere was to weaken barriers to further creole settlement by removing what had been, in effect, buffer states.

Yet if permanent settlement remained limited and tenuous up to the eighteenth century, the effects of European invasion were very evident far beyond the advancing frontiers, and from an early date. The spread of European species changed patterns of agricultural and pastoral activity throughout the continent. The military implications of the horse for warfare between indigenous polities may have been exaggerated, but horses also made more extensive agriculture possible while the effects of European cereals and of sheep, goats and cattle as supplements to diet and foundations for craft manufacturing were everywhere profound, often altering the relative status of distinct social groups within indigenous communities. Disease took a dreadful toll, altering the balance of advantage in favour of the encroaching Europeans. One estimate suggests that the indigenous population of the Americas fell from 50 million on the eve of first contact at the end of the fifteenth century, to as few as 5 million by 1650 (Barraclough 1999). Nor can disease be dismissed as a continuing factor after the initial shock. The disproportionate effects of epidemics of cholera, smallpox and other communicable diseases upon indigenous populations were felt right through to the end of the nineteenth century, if not beyond.

So much for the course, extent and general implications of European settlement. Effective imperial control was another matter. Established quite soon after first contact by the monarchs of Spain and Portugal, direct responsibility for the exploits of subjects in America was not assumed by the French Crown until 1663, while the English continued to operate through an indirect system of sub-contraction to chartered companies, such as the Hudson's Bay Company (1620) or the Massachusetts Bay Company (1628), to which state-like powers were delegated.

The (relative) autonomy of the settler state and the emergence of an American international society

International society in the Americas by the eighteenth century comprised three broad classes of political actor: imperial sovereign states, more or less organized colonial communities, and indigenous polities. Reverberations of European conflicts brought imperial forces into closer and often less than harmonious relations with colonial communities from as early as the mid-seventeenth century, though the heyday of imperial rivalries in the Americas came roughly a century later. The unavoidable conclusion is that European dominion in much of the western hemisphere remained tentative up to the eve of the eighteenth century. By the middle of the nineteenth century the position had been transformed. The Louisiana Purchase of 1803, the admission of Texas into the Union in 1845, and the British and Mexican cessions of 1846 and 1848 had produced something close to today's United States of America. To the south, Chilean, Argentine and Brazilian formal nineteenth-century territorial claims entirely excluded indigenous polities. However, even after the decisive southern conquests of 1878–82, the effective frontiers of Chile and Argentina ran from Puerto Montt on the Pacific, across the Andes, and along the Rio Negro to Bahia Blanca, the only substantial settlement beyond this line being the Welsh nationalist enclave of Chubut.

In part, all this was legal and cartographic prestidigitation. Not all indigenous peoples felt bound by creole maps, let alone law; nor were they even aware of their implications, any more than they fully understood the treaties they had signed. A faux-antique map of early nineteenth-century North America in an historical atlas published in the USA as recently as 2004 blithely designates a great swath of the Mid-West, from today's Montana and the Dakotas in the north to Oklahoma in the south, as 'unorganized territories' (*National Geographic* 2004: 70). The same publication traces 'the spread of religion' (ibid.: 72), as though there had been none before the arrival of Christians, Jews and Hindus. The Lombardis' otherwise excellent atlas of Latin American history acknowledges no surviving indigenous zone in the southern cone by as early as 1830, nor any survival of indigenous peoples as a distinct element in the population or in its analysis of contemporary economy and society (Lombardi and Lombardi 1983: 49, 86–91, 92–104). First Nations are just not there.

It is clear that the creation of present-day Canada, the USA, Mexico, Brazil, Argentina and Chile all depended substantially on the conquest of persistently autonomous indigenous polities over a period of close to 400 years, starting with first contacts and effectively terminating (except in Amazonia) in the final defeat of the Mapuche by Chile in the early 1880s, Julio Roca's 'desert campaign' of 1878–79 in Argentina, and the December 1890 massacre at Wounded Knee, South Dakota. Yet these later events were not the curtain-raisers for an American international society. They resembled rather more the change of state of an existing society from the liquid interaction of a plurality of types of polity to the more uniform relations of a group of modern states. Not for nothing did Arie Kacowicz, an Argentine scholar working in the tradition of Hedley Bull, commence his English School study of South American international society in 1881 (Kacowicz 2005). An American international society was in operation throughout this period, in the sense that relations between polities were governed by norms and institutions. However, a leading purpose of this section has been to establish that there was a decisive phase in this very extended historical process, from the start of which the final outcome was clear, and the sustainability of a society comprising heterogeneous political forms progressively called into doubt. This coincided roughly with a long eighteenth century stretching from the near-contemporaneous Bourbon assumption of responsibility for New France, the British seizure of New York in 1664, and the revived independence of Portugal under the Braganzas, through to the early nineteenth-century independence of much of Spanish America and the migration of the Portuguese court. It was after 1660 that a determined and demographically driven creole break-out from the original European bridgeheads of the sixteenth and seventeenth centuries got under way. It is therefore to the period from 1660 to 1820 that the remainder of this chapter is devoted, and to a comparison of three contested zones: the far south, the south-west Brazilian interior, and the north-east forests of trans-Appalachian North America.

The far south

Spain inherited an Inca tribute empire that petered out somewhere between the present-day Chilean capital of Santiago, founded in 1540, and the Bío-Bío valley, some 400 km further south, in the area that became known as Araucania. The native Picunche in this border region were effectively subdued by 1541, though Spanish demands for foodstuffs and labour prompted some continued resistance and southward flight, which the Spanish hastened to cut off after their pioneering expedition had been reinforced from Peru. Thereafter, a regime of crop levies and labour drafts operated initially within this zone. Following a major revolt in 1598 in which thousands of Spanish men and women were taken captive, the Spanish distinguished between the more biddable Picunche and those peoples living to the south of the Maule and Bío-Bío rivers, who were finally to coalesce into the Mapuche nation but might better be referred to for now as Araucanians or by Boccara's term, derived from the documents: '*los reche*' or 'true men', 'sin Rey, sin fe, sin ley' (without king, faith or law) (Jones 1999: 148; Boccara 1999: 426). In referring to them collectively as Araucanians, no more is intended than to indicate the mutual intelligibility of a multiplicity of culturally related but politically distinct groups.

By a treaty of 1641, confirmed in 1655, the Spanish formally accepted the continuing political independence of those living beyond the Bío-Bío. By this time a more or less stable frontier ran westward from Buenos Aires through San Luis and Mendoza toward Santiago, the only exception being the lands stretching southwards to Concepción to the west of the Andes constituting an exceptional salient. Cross-frontier raids in both directions were common over the next 150 years: indigenous groups riding north and east for women and booty; the Spanish, south, for labour and to administer punishment for northward incursions. General Araucanian uprisings took place in 1723 and 1766. However, the first of these led to the institution, by the Spanish, of a system of *parlamentos*, or regular meetings for the purpose of negotiation, in which thousands of indigenous people might take part, the costs being borne largely by the Spanish. This innovative form of diplomacy facilitated a long, if uneasy, peace on the western side of the mountains from 1773 to 1859, punctuated only by Mapuche support for the Spanish royalists against rebellious creoles during the War of Independence (Brand 1941: 19). However, it also allowed the Mapuche to direct their attention to raiding on the eastern slopes of the Andes, sometimes in concert with the Tehuelche, who predominated in southern Argentina (Jones 1999: 157–59).

The *dénouement* of this process falls beyond the strict time span of this volume. Though much of the world was finding itself ever more obliged to play by European rules from the end of the eighteenth century, no one had told the Mapuche. Araucanian independence continued to be formally recognized by republican Chile until the mid-nineteenth century. At that point southward pressure from land-hungry German immigrants began to cause friction over land. This was aggravated by the support and subsequent refuge provided by the denizens of Araucania for dissident Chilean liberals following the civil wars of 1851 and 1859 (Brand 1941: 22; Jones 1999: 176). The Chileans responded by imposing a reservation policy. This provoked a further revolt between 1868 and 1870, which in turn prompted the Chilean state to move its formal frontier south to the Río Malleco. The end came when the Araucanians took advantage of the transfer of Chilean troops to the north during the War of the Pacific (1879–83). Victory achieved, seasoned troops were sent south where they inflicted a final and irrevocable defeat upon the rebels (1881–84). A cholera epidemic in 1886 finished the job. Formal incorporation of the Araucanian lands into the Chilean state followed speedily, in 1887 (Brand 1941: 23).

Four general features of general relevance to International Relations emerge from the historical literature on Araucania. These are the mobility and strategic flexibility of the Araucanian bands, their conduct of diplomacy, their military transformation, and the political consolidation that arose from resistance.

The first is already familiar from sub-Saharan Africa and other areas of European dominion and settlement. Boundaries imposed by the Europeans and adopted by their creole successors were an inconsistent reflection of pre-conquest society and politics. The Spanish reached the far south by two distinct routes, one to the west and the other to the east of the Andes. This was reflected in administrative divisions, with Chile organized as a military captaincy-general and *presidencia* within the Viceroyalty of Peru in 1606, while present-day Argentina was also part of the Viceroyalty of Peru at first, but became a separate Viceroyalty in 1776. Retrospectively, the much-disputed border between independent Chile and Argentina, following the *cordillera*, came to be regarded as a barely surmountable natural barrier to trade, but trade and migration across the mountains was not uncommon before the Spanish conquest.

Kristine L. Jones counsels her readers that '[t]he first task inherent in understanding interethnic reorganization and re-adaptation at the margins of Spanish rule is to challenge [the] north–south perspective' (Jones 1999: 139). Indeed, archaeological evidence suggests that there was a good deal of east–west movement of goods and economies, based on exploitation of resources deriving from a portfolio of ecologies defined by altitude rather than latitude. However, intensification of trade and migration across the Andes in the sixteenth century, resulting in a measure of political consolidation among the Pehuenche people, who were later to form the core of the emergent Mapuche nation, may have originated as a defensive strategy in response to earlier Inca encroachment. Further pressure, as we have already seen, led Araucanians to reach accommodations with the Spanish on the western side of the *cordillera*, while raiding into the easier plains to the east, where their allies and captives included members of Tehuelche bands, many of which lived in relative peace with the Spanish, engaging in a substantial trade (Jones 1999: 154–59; Brand 1941: 24). In sum, flexible movement across the Andes materially assisted resistance against the Spanish.

Diplomacy and negotiation, formalized in the *parlamento* system only after 1723, had been employed by the Spanish in tandem with coercion from the outset. However, the Araucanian peoples and the Spanish both seem to have felt that binding agreements existed to be flouted by the more powerful party. Padden (1957: 110) noted that the early colonial period saw a repeated sequence of indigenous submission followed by well-planned rebellion.

> The Indians were so skilled that the Spanish were never able to tell whether they really meant to keep the peace or not. Either way, there was nothing the Spanish could do but accept, driven as they were by the economic imperative of their way of life. The Indians seem to have understood this perfectly. The granting of supposedly permanent peace

> terms in critical areas as a part of wider strategy became common. Revolts were well planned: the chiefs had the white women divided among themselves in advance …

The confidence underlying what appeared, to the Spanish, to be contemptible duplicity must have been reinforced not only by an ability to retreat into the mountains and beyond them, out of Spanish reach, but also by growing military competence. Araucanians did not initially fight on horseback, and the initial impact of the horse was probably to assist trans-Andean trade and migration, and facilitate alliance formation and the concentration of forces (Gregson 1969: 36–37). Beyond the *cordillera*, the Tehuelche did not exploit the wild herds roaming the *pampa* much before 1700 (Gregson 1969: 34). Initially, the Araucanians carefully chose battle sites that gave them an advantage by nullifying the Spanish use of horses (Padden 1957: 112). Yet, by the late sixteenth century they had developed a light cavalry more mobile and effective than that of the Spanish (Padden 1957: 113; Gregson 1969: 37).

Military competence was transformed not only by the spread of horses but by a more general and profound cultural transformation of the Araucanian economy prompted by the adoption of European livestock species. This led to a change in gender relations, with women taking exclusive responsibility for horticulture and textile crafts while men engaged in commerce and raiding across a wider territory than before. Spatial extension of external trade, especially in hand-woven *ponchos* (capes), provided the basis for the Araucanian war economy in the shape of metal weapons, better-quality mounts and surplus males (Boccara 1999).

East–west strategic flexibility, the judicious alternation of diplomacy and force, and military transformation would probably not have sufficed to prolong Araucanian resistance had it not been for political consolidation. A scrupulous economy in use of the term 'Mapuche' has been observed thus far, because there was no self-conscious Mapuche nation or *ethne* much before the late eighteenth century. The term first appears in the written record in 1760 and began to be used for self-description around the same time. Padden, writing of the seventeenth century, found that resistance to the Spanish was organized in three, later four, north-south alliances of Araucanian peoples, called *butanmapos* (Padden) or *futamapus* (Boccara), with specialist officials fulfilling the functions of general command, field command, and envoy or religious leader (Padden 1957: 118–19; Boccara 1999: 426ff). The literal meaning of this term was 'great land' and it appears to have emerged in the early seventeenth century as the coming together or supersession of an earlier social form, the *ayllarehue* or 'new true place', a temporary alliance or non-aggression pact with no unified command. Both the *ayllarehue* and the more general *futamapu* bound together many *lebos* or bands, and the effect of prolonged hostilities against the Spanish together with participation through a *futamapu* leader in the *parlamentos*, led to the consolidation into a collective political identity of forms originally adopted for ad hoc military purposes.

South-west Brazil

In Brazil, as elsewhere in the Americas, the eighteenth century was pivotal. This was largely because of the vicissitudes of dynastic rivalries in Iberia itself. There, extinction of the ruling house of Aviz by the death of King Sebastian of Portugal in 1578 at the Battle of Alcazarquivir led to a succession crisis that provided an opportunity for Philip II of Spain to seize control, declaring himself Philip III of Portugal. Under its new Habsburg overlord, Portugal retained internal autonomy but lost control of foreign policy and saw its overseas possessions relatively neglected. The ongoing struggle between Spain and its rebellious subjects in the Netherlands spread to the New World, where many parts of the Brazilian coast were occupied by the Dutch during the second quarter of the seventeenth century. Elsewhere, the French also encroached. The overbearing posture adopted toward his Portuguese subjects by Philip III of Spain led to a successful revolt following the installation of the Duke of Braganza as king of Portugal in 1640. The Spanish did not readily accept defeat, and a war of independence continued into the 1660s. Only as the threat in Iberia was overcome were the Portuguese able to retrieve the position in the Americas, ousting the Dutch from Brazil in 1654 and creating conditions that would allow the long move into the interior to begin in earnest.

In the south, where the Dutch had been held off, Portuguese slave-raiders had already begun to operate out of São Paulo, but it was not until 1693 (Minas Gerais) and 1718–23 (Goiás) that mineral discoveries sparked a rush of Portuguese with their African slaves, encouraged by construction of the *estrada real* (royal highway) connecting Rio to mining centres in the interior. This speedily put paid to the political autonomy of the Goiases in their traditional homeland (Karasch 2005: 466), but many Goiases and Crixá fled into the surrounding forests and continued to resist from their relative safety. The Crown was anxious to close the frontier yet at the same time feared a leakage of bullion down the Paraná valley to the Río de la Plata. To this end, slaving expeditions (*bandeiras*) into the region east of the new mines were forbidden. However, *bandeiras*, albeit on a smaller scale, continued to press east, now less for slaves than in the hope of further lucky strikes, and were active throughout the eighteenth century across the Portuguese realm (Langfur 2005: 460).

Mary Karasch (2005) questions Portuguese claims to have conquered indigenous peoples and closed the frontier. Concentrating on the province of Gioá, she notes at least one group – the Avá-Canoeiro – that was not subjugated until the twentieth century, but devotes most of her attention to three peoples whom the Portuguese *did* claim to have conquered: the Karajá (1775), the Kayapó (1780), and the Xavante (1788). The first of the three suffered the depredations of several *bandeiras* during the seventeenth and early eighteenth centuries, culminating in a 1755 raid that killed many and enslaved others. In an attempt to subdue them, an officially sanctioned expedition, more properly referred to as an *entrada* than a *bandeira*, was sent in 1774 with instructions to avoid violence. The Karajá submitted to Portuguese authority voluntarily after extended negotiations, but their subsequent experience of unfair exchange, missionary life and epidemics led to a revolt in 1813 (Karasch 2005: 480). Like the Karajá, the Kayapó were raided by *bandeiras*, notably in 1741 and 1742. They retaliated, and a pattern of reciprocal violence was established, ending only when the Portuguese decided to buy them off. Once again, the decision to submit turned out badly as the Kayapó found themselves forced to labour for the Portuguese and were stricken by disease. Those who survived fled into the Araguaia valley in the early nineteenth century and continued an intermittent guerrilla war against incoming creole and European settlers right up to the 1880s, even though their numbers were decimated by an official punitive expedition in 1824. A third group, the Xavante, reached a negotiated settlement with the Portuguese in 1788 and were resettled on lands in Carretão that had been promised them by the governor of the province. There – weakened by epidemics and suffering further losses in reprisals following the 1813 revolt – their numbers had dwindled to 227 by 1819 (Karasch 2005: 484). Here, as elsewhere, the process represented by Europeans as conquest extended over many decades and had rather more the character of an international society, albeit asymmetric and unsustainable in the long run.

Similar contrasts may be drawn, throughout the *sertão*, between confident declarations of Portuguese victory and persistent indigenous resistance. Langfur notes that the Bororo of Mato Grosso, regarded as a serious military threat to Cuiabá, the state capital, as late as the 1880s, were first said to have been subjugated more than 150 years before, in 1730 (Langfur 1999: 881). Always, however, the tendency was in the same direction, toward progressive marginalization of the indigenous peoples of the interior.

Standing back from the band-by-band narrative of resistance and accommodation to the creole invasion, it is possible to see some similarities with experience in the far south. Once again the mobility of indigenous peoples is striking, and reminiscent of the hordes that dominated Central Asia from the thirteenth to the fifteenth centuries of the common era, or the militarized states of the nineteenth-century southern African Mfecane (Omer-Cooper 1966). Sometimes as refugees, sometimes seeking the protection of Jesuit or other mission settlements, sometimes by voluntary agreement with the state, bands moved to new areas. The readiness with which they sometimes did this may reflect a sense that the lands from which they were being removed did not constitute a traditional homeland; it may even indicate a more fundamental difference between European and indigenous attitudes to territoriality. From the sixteenth century onwards *bandeirantes* plunged into the *sertão* like so many cue balls, cannoning the reds in every direction in their urgent and careless desire to pocket the colours. This is not only a sharp reminder to beware the ethnographers' lust for virgin peoples but also helps to explain a recurrent and important feature of American international society, namely a complex mixing of populations on both sides of the advancing frontier that permitted quite distinct cultural and political divisions to continue even when a supposedly colonial *bandeira* might consist largely of free blacks and

Indians while its victims comprised a *mélange* of creole renegades, escaped slaves and aboriginals. The question this poses to traditional English School accounts is very like E.H. Carr's question to the status quo powers of the 1920s. Did not their readiness to assume that homogeneity of political form was an essential precondition of international order rather too conveniently suit their still imperialist retrospective gaze? Were they not deceived in much the same way as Britain, France and the USA had been when they convinced themselves that peace was in the general interest? The length of time in which a heterogeneous international society endured in the Americas in spite of technological asymmetry and demographic imbalance may be thought testimony to the possibility that such systems may not in themselves be unstable. In the long run, homogeneity may be no more essential for the orderly operation of an international society than balance is for a stable state system (Wohlforth *et al.* 2007a; Wohlforth *et al.* 2007b).

The process of miscegenation between peoples within this society must have been accompanied by some transfer of technology, including military technology, but there is scant evidence of this in the Brazilian expansion, a process generally characterized by smaller raiding parties and indigenous polities than in either the far south or the North American Great Lakes region. One hint is that within a secular demographic collapse, periods of relative peace and trade with the creoles, as sometimes under Jesuit protection (Reeve 1994: 126), allowed populations to recover and become wealthier, an essential precondition for renewed resistance. However, miscegenation also facilitated styles of diplomacy in which gifts, feigned submission, and the employment of bilingual and mixed-race former captives were common. This may be one place to look for the antecedents of green-on-blue attacks.

The northern plains and forests

A history of population decline and radical social disruption even prior to direct contact with Europeans, evident in South America, was if anything more pronounced, and of earlier inception in trans-Appalachian North America. A populous and extensive pre-Columbian Mississippi culture had flourished for half a millennium, up to 1400. Prior to European contact, for obscure reasons, the northern heartland of this culture – based in that stretch of the Mississippi valley between its confluences with the Missouri and the Ohio – had collapsed. Some 200 years later, across a large expanse of the north-central plains, indigenous bands racked by newly introduced infectious diseases were attempting to rebuild their lives while simultaneously dealing with the consequences of a complex triple frontier of advancing French, English and Spanish settlement.

On a distant periphery of this ancient culture, in a great triangle of forested lands several hundred miles to the north-east, bounded by the modern cities of Cleveland, New York and Montreal, a third story of the long eighteenth century unfolded as the British and the French began to break out from their initial tentative bridgeheads in the valleys of the St Lawrence, the Potomac, the Hudson and the Connecticut and James Rivers after English displacement of the Dutch in 1664.

In the eye of the storm stood the Iroquois. Certainly not a consolidated polity, the union achieved by the so-called Five Nations – Mohawks, Oneidas, Onondagas, Cayugas and Senecas – is more accurately described as a non-aggression pact than a confederation and bears some family resemblance to the inchoate Mapuche polity of the same era. They had already fought a succession of wars against the French and their Huron allies, which used to be regarded as having been motivated primarily by competition for pelts for sale to the Europeans (Hunt 1940). Naroll (1969) wrote specifically about the fourth of these (1657–67), arguing that it was not just about the fur trade but stemmed from a mutual lack of understanding between very different cultures. The war started from a relatively small incident, but by 1660 the Iroquois had cut off the fur trade and besieged Villemarie (Montreal). These initial successes were reversed as a coalition of indigenous peoples formed against the Iroquois. The strain of a war on two fronts, coinciding with an epidemic, put them on the back foot. Then, in 1663, the French government in Paris formally adopted the Canadian settlements and sent a regiment of 1,000 regulars who mounted a punitive expedition that killed few, but destroyed Iroquois crops and villages. In 1667 the Iroquois sued for peace, accepting all French conditions.

The year 1675 saw peace between the Mohawk and Mohicans, to their north, and between others of the Five Nations and the Susquehannock, to the south. With their northern and southern frontiers secure, the Iroquois attacked groups to their west. At the same time, the British and French began to clash over the lucrative North American fur trade. The British had acquired New Amsterdam in 1664. Ten years later a new governor set about enlisting the Five Nations as his auxiliaries in a general pacification of indigenous peoples (Richter 1983: 545) and a challenge to French suzerainty.

To counter the British threat, the French started to build forts in the Illinois country. Finding so many furs being taken to Albany by the Iroquois, the French decided to crush them and mounted repeated invasions, spoiling crops and villages, from 1684 onwards. French expeditionary forces of up to 2,000 took part in these operations. British responses to French incursions were largely ineffective and poorly coordinated with the Iroquois during the War of the League of Augsburg (1688–97), generally known in Anglophone North America as King William's War. After the Treaty of Ryswick had brought hostilities to an end in Europe, an unresolved dispute between the French and the British in North America led to continued fighting there, the former demanding a separate peace with the Iroquois while the latter insisted on their right to negotiate on behalf of their allies. Not until 1701 was a settlement agreed that included a general peace between the indigenous polities formerly allied to both European empires. This did not bring peace to the Great Lakes area. The Iroquois reverted to the traditional form of mourning war, in which men were encouraged to replace individuals in their community who had died by raiding, so that *killing* the enemy was close to counterproductive. From the 1720s, avoiding any breach of the 1701 treaty by avoiding war with its signatories, the Iroquois continued raiding Flathead groups on the western frontiers of Virginia and the Carolinas for the remainder of the eighteenth century.

As in the far south and the Brazilian *sertão*, so in the forests and lakes of trans-Appalachia, the long eighteenth century witnessed the terminal struggles of indigenous peoples, already transformed socially by their direct and indirect contacts with Europeans and creoles, as they deployed negotiation and force, by turns, in their attempt to preserve autonomous political forms which, by the end, had become anything but traditional. A profusion of treaties between Eastern seaboard colonies and native polities raises the question of whether their legal status was appreciated in equal manner and measure by creole and native signatories. Similar ambiguities surround the despatch of envoys by the Cherokees to the Spanish and French, and the conferral of European titles by colonists and imperial plenipotentiaries on native leaders. May one justly speak here of diplomacy, balancing and mutual recognition?

It would be wrong to say that the fur trade was the sole or chief cause of war between the Iroquois and their neighbours. Naroll finds that they had access to enough beaver pelts for their needs without having to fight for them; possession of material goods did not confer status as it did for Europeans; there was no mention of the trade in the 1655 treaty with the French (Naroll 1969: 59–61). However, at this point, the comparative method adopted here poses questions. The nature of war in the north changed markedly in the course of the sustained conflicts of the eighteenth century, from an institution primarily concerned with blood feuding and the replenishment of population, to a much bloodier contest over territory.

Naroll's view of the matter is exposed as too static by subsequent scholarship, and Schlesier (1975) offers a more dynamic view. The French found it puzzling that their local allies were satisfied to break off engagements after inflicting light casualties and taking a few, carefully chosen prisoners. They did not readily grasp the concept of a mourning war, in which capture, not killing, was the primary objective of battle.

What happened toward the end of the seventeenth century was that population decline made replacement a more urgent and larger-scale enterprise, while the supply of more deadly weapons and the involvement of European troops raised the level of casualties, and the consequent demand for replacements, to the point where the ritualistic form of mourning warfare collapsed (Richter 1983: 544–51). 'French attempts to apply European ideas of mass destruction in their dealings with their recklessly chosen enemy, the Iroquois', so Schlesier claims, 'were exactly what created the power and the "menace" of the Iroquois about which French colonials later lamented very bitterly' (Schlesier 1975: 131–32). In the course of this transformation of the nature of warfare, North American Indians came to think of themselves as one people, a change of political identity reminiscent of the contemporaneous ethnogenesis of the Mapuche, far to the south, fostered in this instance by prophets whose words soared across tribal divisions during the second half of the eighteenth century (Dowd 1991). It is hard to imagine that such transformations can have taken place without changes in gender roles and the significance of material goods of the sort that are seen elsewhere. In short, the insistence of Naroll on the persistence of an egalitarian attitude to material possessions gives grounds for suspicion.

In respect of negotiated agreements, the pattern is very much the same as in the far south and the Brazilian *sertão*. Indigenous groups often made solemn compacts only to break them without warning soon afterwards. Sometimes this was because they did not feel bound by contracts concluded under duress; sometimes the problem was more fundamental, and stemmed from European failure to understand the nature of their negotiating partner. The Iroquois confederacy was a pact designed to regulate blood-feuding between its members. The French at times regarded it as a true federation, able to control the foreign policy of all its component peoples. They therefore felt aggrieved when one or other of the Five Nations took independent action. The Mohawks, as one of those component groups, saw no inconsistency in negotiating with the French while simultaneously attacking their indigenous allies. The French, by contrast, thought this deceitful and, by and by, each side became convinced of the bad faith of the other (Naroll 1969: 60, 69, 71). For all this, the confederation represented one step in a process of political integration and the prophecy movement described by Dowd another, each of which must surely owe something to sustained struggle with culturally discrete enemies.

Conclusion

It is tempting to try to gather together observations from this too brief survey of American ethno-historical literature and generalize from it using standard categories of a modernist international society approach to international relations: diplomacy, law, international organization, war, balancing and hegemony, and trade. An approach of this kind would rapidly encounter two great obstacles. The first is that the range of experience, even from a selective sample of the literature, is such as to defy easy generalization, so that the most that can be hoped for is to identify an agenda. The second is that the categories of European and creole experience cannot readily be applied transculturally, as the first generation of English School scholars were inclined to do. To attempt this is to deny the lived experience of many of the participants in the process under investigation.

There are two reasons for this. The first is that understandings of particular forms of engagement differed. It is evident from all the cases considered here that war and diplomacy were very differently practised and understood by indigenous and creole peoples. However, a second and more fundamental problem is that even this way of stating the problem imposes alien categories on indigenous experience. Trade, for example, can hardly be separated from diplomacy, or indeed from warfare, in a world of semi-nomadic peoples where gift-giving and exchange remained – as they had once been in Eurasia (Allsen 1997) – a central expression of relative power and status, while warfare often took the form of raiding for captives and livestock, and long-distance trading expeditions might at any point have to take up arms. The period dealt with here is precisely that in which the foundations of the Euro-American liberal world view were laid. One of its pillars was the naturalization of a very particular form of social analysis in which the economic and the political were positively separated and normatively divorced. To examine the ethno-history of this epoch is to be made aware of the strangeness and specificity of this worldview, and its ideological force.

Just as trade defies any easy separation from diplomacy or warfare, so group identities – the ontological basis of International Relations – resist easy identification or classification. From the vantage point of the present day it is customary to tell a story of demographic collapse, imperial competition, the progressive emergence of creole nations, a long period of subjugation of indigenous peoples, and their recent resurgence. Yet, examples of indigenous ethnogenesis and state formation are observable throughout the hemisphere during the long eighteenth century. It was not only the Iroquois and the Mapuche who consolidated concurrently with their creole neighbours, but also the Catawbas and the Cherokee (Salisbury 1996: 432–33). Self-identification as Mapuche, Aymara or Sioux in the early twenty-first century is a cultural gesture lacking secure genetic certification. Its strategic function is to enable marginalized groups to aspire to legal and political parity with their oppressors in a contemporary world, where orderly international society is believed to be premised on homogeneity of its members. Polities were not firmly rooted in specific territories in Brazil or the northern forests, where disruption arising from multiple European frontiers was already underway by the seventeenth century. Perhaps they never had been. Even in the far south, where the attachment to territory was more stable, trans-Andean raiding and migration were stimulated by war against the Spanish and led to growing genetic heterogeneity of the population at the very time when it was becoming more united politically.

If it is mistaken to attach too much importance to primordial territories, it is also quite wrong to imagine that indigenous polities were genetically homogeneous. How could this be the case in a world where raiding was so common, creating cultures that were primed by custom, even before first contact, to receive and integrate European captives and renegades? Nor did the forces they faced consist of pure-blood creoles, still less Europeans. Brazilian *bandeirantes* were more often black or Indian than creole, though generally led by a Portuguese. Mixed-race forces were the rule rather than the exception in the North American wars of the eighteenth century, while the

Argentine forces deployed by Rosas and Roca against the southern nomads included many Afro-American and mixed-race soldiers.

In a world where war was conducted between racially mixed forces, diplomacy would have been close to impossible without Janus-like figures such as the French *coureurs du bois*, English frontiersmen and returning captives. 'One explanation for the success of … *bandeiras* in negotiating peace via gift giving', Karasch (2005: 491) explains, was that 'each "conquest" utilized indigenous "ambassadors" as interpreters, warriors, and informants in subsequent "conquests"'.

In short, the history that has been reviewed here is one in which certain cultures and political forms proved unsustainable, in spite of fierce resistance, but where the identification of these cultures and polities with specific nations is a far harder trick to pull off, and the safest course may be to resist the easy view, of progressive invasion and annihilation, and think instead of a prolonged and norm-governed continental civil war out of which emerged discrete American *nations* based on miscegenation and its denial, characteristic American *polities* defined by creole foundational myths, and a distinctive style of *international relations* based on delusions of escape from the superstitious, corrupt and hereditary systems of the Old World. Above all, the American eighteenth century offers a salutary warning against the synchronic vision of Bull, Watson and their contemporaries, in which the past converges on an idealized set of modern norms and institutions that provides the paradigm of international society. The international society described here was one of constant change, in which norms and institutions were always provisional and subject to the play of power. The outcome was not a new Europe, but a distinctively American international society in the nineteenth and twentieth centuries, marked off from Europe by distinctive practice and ideas in the fields of international law, international economic relations, public violence and political exile (Jones 2007b; Snajder and Roniger 2009; Roniger *et al.* 2012).

Experience varied widely, but it is possible to put forward a few hypotheses about the ways in which international society in the Americas during the long eighteenth century prefigured its later forms of institutionalization following the final military triumph of independent creole states over their European and indigenous competitors. One point of departure, following the demographic catastrophe that befell the American nations as a consequence of first contact with Europeans, is to be found in the dichotomy between imperial benevolence and creole ruthlessness. Whether it be Italian Jesuits providing havens in the *sertão* and the Amazon, British officials vainly holding back the westward drift of creole settlement after the Seven Years War, or the Portuguese Crown restricting the range of operations of the *bandeiras* and urging diplomacy on the leaders of their pacification expeditions or *entradas*, an almost universal tension is evident between the posture of European elites and creoles fully committed to the New World. When the latter came to power, following wars of independence, they rapidly jettisoned the tolerance of constitutional variety and historic rights to which the composite monarchies of early modern Europe has, albeit imperfectly, aspired. In only the baldest of such statements, soon after assuming responsibility for the French colonies in North America in 1663, a French official advised that 'it is [the king's] intention that the officers, soldiers and all his other subjects treat the Indians with kindness, justice and equity, without doing them any injury or violence, and that no one encroach upon the lands that they occupy under the pretext that they are better or more convenient to the French' (quoted in Naroll 1969: 65). It became increasingly evident as time went by that creoles would win out against their imperial overlords, and the alignment of indigenous peoples with royalists in many parts of the continent during the wars of independence was a doomed strategy. 'The news that Britain had conceded defeat and was leaving them to deal with the Americans on their own', Neal Salisbury (1996: 453) reports, 'was met with disbelief and then concern and resentment by Indians everywhere'. With the British went the 1763 Proclamation Line reserving for the Crown the right to alienate land between the Appalachians and the Mississippi.

Because of the genetic mêlée surrounding the frontier, maintenance of strict cultural and military distinctions required creoles to deny the extent of miscegenation through a combination of ethnic cleansing, statistical manipulation, promotion of European immigration and resolute self-denial. Much the same was true of military culture, where a reality of genocidal attrition was transformed by the deployment of uplifting narratives of the armed forces as *desarollista* nation builders, blazing trails, building real and figurative bridges, and generally making a world safe for settlement, or else as guardians of a constitution that could often be preserved only in the breach (Nunn 1983; Larrain 2006). Alternatively, as in the USA and Canada, traditions of high-intensity expeditionary conflict or benevolent peace keeping and humanitarian intervention could be substituted for the less attractive reality of *petite guerre* (Grenier 2005).

Add to this a tradition of diplomacy in which force is never very far from the negotiating table. The long history of pacts made under duress by all parties and unashamedly broken by each at the first sign of opportunity makes it easier to understand the seeming contradiction of prevalent sabre rattling and infrequent warfare that has distinguished Latin America from other regions over the last two centuries (Kacowicz 1998; Mares 2001); easier, too, to understand how necessary it has been for states throughout the continent to engage in a Nietzschean forgetting of history and embrace quasi-disciplines of International Law and International Relations that promise to occlude their own formative experience. Were it not for this plugging-up of every cultural orifice, it would have been impossible for creole self-images of American liberalism, toleration, constitutionality, racial tolerance and love of peace to have been maintained even as unattainable ideals. It is hardly to be wondered that this episode of ideological flatulence should have ended with a monumental historiographic fart, ripping through the hemisphere, as self-defined First Nations have set about retrieving invented pasts for present purposes (Harris 1995). What is interesting, though hardly unsurprising, is that it should have taken so long for International Relations, a field ostensibly concerned to whisper truth to power, to detect the stench and realize its own complicity.

Notes

1 The term creole is used throughout as an anglicized form of the Spanish '*criollo*' to denote people of predominantly European descent born in the western hemisphere.

2 The tenacity of the Jesuit missions provides a complement to the powerful reminder provided by Janice Thomson, that the modern European state was not without rival organizational forms even in its European heartland. See Thomson, J.E. (1994) *Mercenaries, Pirates and Sovereigns*, Princeton, NJ: Princeton University Press.

7 Europeans, Africans and the Atlantic world, 1450–1850

Joel Quirk and David Richardson

Introduction

Cross-cultural exchanges have historically taken many different forms, yet International Relations (IR) scholars have tended to focus narrowly upon a specific subset of cases associated with the history and theory of European imperialism and colonialism (e.g. Doyle 1986; Barkawi and Laffey 2002). The priority attached to European imperialism has meant that the history of cross-cultural exchange has been most commonly approached in terms of the forcible imposition of European preferences and agendas, with European actors taking advantage of their dominant ideological and material resources in order to manipulate and coerce their weaker counterparts in other parts of the globe (notwithstanding various manifestations of defensive negotiation and contestation). While some historical examples undoubtedly fit within this familiar template – especially from the mid-nineteenth to mid-twentieth centuries – there are also no shortage of historical examples where this default presumption of European dominance has proven misguided and misleading (Hobson 2004: 134–57; Darwin 2008: 93–97, 186–201). As other chapters in this volume can further attest, patterns of cross-cultural exchange between Europeans and peoples in other parts of the globe have been regularly defined by forms of negotiation, limitation and adaptation, rather than unilateral European imposition. In many settings Europeans found it difficult to consistently impose their preferences upon their counterparts elsewhere, and were obliged to develop alternative means of transferring items and ideas across cultural and political divides. These cross-cultural exchanges were not always equally beneficial to all of the parties involved, but they would not have taken place at all if those involved had not expected to benefit in the first place.

This last point is crucial to understanding the history of European incursions into (what became) Atlantic Africa in the early modern era. During the roughly four centuries that preceded the 'Scramble' for Africa in the late nineteenth century, European traders and officials of many nationalities developed long-standing relationships with many of their coastal counterparts in Western Africa. As we shall see, a large portion of these links were based upon slave trading, especially from the mid-seventeenth to early nineteenth centuries. Cross-cultural slave trading was not only defined by tremendous levels of suffering and death – both in Africa and the Americas – but has also been widely identified as a key contributing factor when it comes to continuing problems facing many contemporary African states and their citizens. Under these circumstances, it is very tempting to explain African involvement in slave trading as a (by)product of European coercion and manipulation, yet this formula ultimately runs against the grain of a now significant body of research which suggests that the slave trade was often not so much imposed as negotiated.

This chapter aims both to describe and explain some of the defining characteristics of patterns of early modern cross-cultural exchange between Europeans and Africans. In pursuit of this goal, we have divided the chapter into five sections. In the first section, we briefly reflect upon the relative neglect of pre-colonial Atlantic Africa within existing IR scholarship. The second section then goes on to consider other approaches to our chosen topic within the social sciences more broadly. Our main argument in these two sections is that many of the distinctive features of pre-colonial exchanges in Atlantic Africa have regularly been overlooked, either through a lack of interest or expertise, and/or as a consequence of a widespread tendency to define the last five centuries in terms of a lasting pattern of European domination and African dependency. In section three, we review some of the main cultural, political and economic attributes of Atlantic Africa in the early modern era. In addition to offering a brief introduction to the region, this section explores the link between slavery and larger social formations.

In section four, we look at the early history of cross-cultural exchanges between Europeans and Atlantic Africans. We pay particular attention to the history of the Portuguese in Angola, which was the one area where a major colonial footprint emerged before the nineteenth century. This history is important, because it offers insights into the capacity of Europeans to impose their preferences upon their African counterparts. One of the key themes to emerge here is the disjunction between the individual gains that accrued to both European and African slave traders and the larger collective costs of slave trading for peoples in Africa. Finally, we review a number of cross-cultural institutions and influences that characterized European dealings in Atlantic Africa more broadly. Transatlantic slavery would not have persisted for such a long period of time, or reached so many parts of the Atlantic coast, without a series of cross-cultural institutional arrangements that helped to standardize how trade took place, what units of exchange were used, and how contracts and credits were enforced. As we shall see, this regularly involved the repurposing of existing institutions and expectations to a variety of new ends, with Europeans commonly adhering to protocols and preferences established by their African counterparts.

Africa and IR

International Relations scholars have rarely given sustained or significant attention to African experiences or perspectives, either historical or contemporary (Smith 2009; Nkiwane 2001; Dunn and Shaw 2001). This situation can be partly attributed to the longstanding Eurocentrism which has largely defined historical research in IR circles, and partly attributed to the marginal status generally accorded contemporary Africa within IR. The most influential theorists and analysts of the actions and outlooks of states in Africa have not been IR scholars, but instead hail from other disciplines (e.g. Mamdani 1996; Mbembe 2001; Bayart 2009; Young 1994; Comaroff and Comaroff 2007). To the extent that Africa has been considered within IR circles, the focus has almost entirely been on fairly recent developments, such as colonial rule and the causes of decolonization (Crawford 2002; Philpott 2001; Jackson 1990) or more recent post-colonial global and regional relationships (Taylor and Williams 2004; Taylor 2010; Cornelissen *et al.* 2011).

As a consequence of this contemporary focus, the pre-colonial period of Euro-African exchanges has generally been treated as a 'backstory', which receives (at most) cursory consideration before moving on to the main event: colonialism and/or post-colonial politics. A good example of this dynamic is Christopher Clapham's influential book *Africa and the International System* (1996: 28–30), which allocates three pages to Africa prior to the imposition of the 'colonial grid', and then quickly moves on to more pressing colonial and post-colonial concerns. On the rare occasions where historical developments prior to the late nineteenth century do receive sustained consideration, such as the work of Siba N'Zatioula Grovogui (1996), they nonetheless tend to be coupled with the later history of conquest and colonization, which has the indirect effect of framing the early modern as a benchmark, or precursor, the main narrative purpose of which is to help better understand the causes, characteristics and consequences of colonialism and its more recent legacies.

There have been, however, a number of occasions where Euro-African relations have been touched upon within larger historical discussions. International Relations scholars have long used historical analysis as a platform for advancing various theoretical arguments and agendas (Quirk 2008). The main focus here has been developments within Europe, with particular attention being paid to relations between Great Powers and transformations in the foundations of political authority (e.g. Gilpin 1981; Teschke 2003). Developments outside Europe have sometimes featured in these larger historical discussions, but they tend to be parochially assessed in terms of their impact upon 'core' European actions, outlooks and institutions (e.g. Philpott 2001; Reus-Smit 1999; Krasner 1999). One prominent example of this larger dynamic comes from Adam Watson (1992: 265), whose book on *The Evolution of International Society* includes several chapters on European expansion, but only briefly touches on Africa as part of a larger survey of how nineteenth-century Europeans 'brought the whole world for the first time into a single net of economic and strategic relations'. A comparable lack of presence can also be found in Barry Buzan and Richard Little's *International Systems in World History* (2000: 263), wherein Africa receives a handful of cursory references, nearly all of which relate to nineteenth-century expansion.

One of the more sustained examinations of Euro–African relations comes from Hedley Bull (1984b) in his chapter on 'European States and African

Political Communities', in *The Expansion of International Society*. Drawing a sharp divide between pre- and post-1880, this chapter offers a brief sketch of the main political communities found in Africa during the early modern era, followed by a similar overview of trading relationships and diplomatic engagements. While Bull's chapter pays much closer attention to the distinctive features of the early modern period than most, his analysis still suffers from the Eurocentrism that has long plagued IR scholarship.

A good example of Bull's (1984b: 111) perspective is found in the following passage:

> How far did European states and African political communities conduct 'normal' relations with one another before the latter were absorbed by the former? In the long period of contact before the final scramble European states displayed a considerable disposition to acknowledge that African political communities had rights of independent existence. There cannot have been anything 'normal' in the experience of European states, or comparable to their relationships within Christendom or Europe, about contacts with societies that were pre-literate, pagan, and in some cases stateless.

When the final outcome – colonial conquest – is known, it is difficult to avoid the temptation to treat everything that came prior to conflict as inevitably leading to this European-dominated endpoint. Taken to its logical conclusion, this approach essentially suggests that Europeans could have colonized Africa earlier, but instead exercised 'restraint', or a 'lack of interest', thereby framing the terms and timing of colonization in terms of inclination, rather than capacity. This is reflected in Bull's reference to a 'disposition to acknowledge', which treats the question of the 'independent existence' of African rulers as something that Europeans had within their discretionary power to grant. As we shall see below, Europeans typically lacked the capacity to threaten the independence of most of their counterparts in Africa during the early modern period, so whether or not their independence was formally acknowledged was not a key priority from an African perspective.

Bull also casts his analysis in terms of a series of binary oppositions – Christian versus pagan, state versus political community or even 'stateless', normal versus 'cannot be anything normal' – which ultimately cast the African side of the ledger as a deviant exception. There are numerous problems with this framing. One of the main issues is that Bull presumes that relations between states 'normally' take place within the context of a shared religion, culture and civilization, when the more common historical scenario instead revolves around states and their agents interacting with regional or even global counterparts who have different cultures, institutions, forms of political legitimation and/or religion. The 'Indian Ocean World', of which eastern Africa was part, offers a useful counterexample here, with longstanding networks spanning the Indian sub-continent, South-East Asia, Arabia and East Africa operating in the absence of a shared 'civilization' (Chaudhuri 1985;

Frank 1998). Rulers in the Middle East, East Asia, Central Asia and many other regions were also familiar with this dynamic, as they were routinely obliged to engage with a multiplicity of political institutions and cultural forms (Darwin 2008; Cooper and Burbank 2011; Göl, this volume). Framed in these terms, Euro–African relations can be viewed as a representative example of a much larger pattern, rather than an exception to the 'norm' of Europe/Christendom.

Like their counterparts elsewhere on the globe, European and Africans developed a variety of institutions and arrangements that helped to facilitate and standardize long-term exchanges across cultural and political lines. While violence was by no means absent from the equation (and when is it?), the absence of a shared culture or civilization did not prove a fundamental impediment during the roughly four centuries of contact that preceded the Scramble for Africa. As we shall see, overlapping economic interests amongst various Africans and Europeans contributed to the emergence of a limited, but still identifiable international order, or more precisely, a series of overlapping international orders; however, these orders were mostly confined to protocols and mechanisms for regulating and regularizing cross-cultural trade.

Europeans and Africans: from chains to bonds?

In the absence of a significant IR literature on our chosen topic, it is necessary to expand our horizons to the social sciences more broadly. It is here that it becomes necessary to engage with an influential strand of thought which links numerous contemporary problems in Africa to a long-term historical pattern of dependency, exploitation and underdevelopment. From this standpoint, there is not necessarily a substantive difference between the pre-colonial, colonial and post-colonial periods, since European domination and exploitation has been a constant and enduring feature of Euro–African relations throughout. Three main claims tend to be associated with this strand of thought. First, we have the idea that many problems facing Africa can be primarily attributed to the adverse impact of external actors and influences, most notably European and later Western imperialism and colonialism. Second, we have the idea that these external actors have exercised dominance over African affairs for an extended time period, thereby establishing an underlying continuity of negative influence that binds together centuries of history under a common rubric of dependency. This negative influence is most commonly understood in terms of political manipulation and economic exploitation, but some accounts also include a cultural component, with 'corrupted' or 'immoral' behaviours within Africa being traced to European intrusions. Finally, we have the idea that this continuous process of external exploitation has produced various forms of unjust enrichment, which helps to explain why Europe (or the West in general) has become rich while Africa remains (or has become) poor.

Our particular concern in this section is to assess the extent to which pre-colonial history – or at least the pre-colonial history of transatlantic slavery in

Atlantic Africa – can be comfortably situated within this model of dependency. This question extends to both transatlantic as a specific period, and to the place of transatlantic slavery within a larger history genealogy. In the case of the former, historians such as Walter Rodney (1972) and Joseph Inikori (1977, 2002) have argued that Europeans dominated and duped their African counterparts during the era of the transatlantic slave trade, thereby enriching themselves at the cost of African lives and African development. More generally, slave trading has been further depicted as the first phase in an enduring pattern of exploitation and marginalization that has extended through colonial rule to the later neo-imperialism that followed decolonization (e.g. Howard-Hassmann 2008; Nunn 2007, 2010; Diène 2001). Transatlantic slavery occupies a pivotal role within this larger historical genealogy, as it establishes an extended timeline which suggests that African dependency began centuries prior to colonization. The vast majority of Africa – Atlantic or otherwise – was not colonized by Europeans until the late nineteenth century. Adding transatlantic slavery pushes dependency back as far as the mid-fifteenth century.

The fundamental problem with this historical genealogy is that it places too much emphasis on continuity and not enough emphasis on change. Important differences between the early modern period and subsequent events have consequentially been overlooked or sidestepped. However much slave trading raises difficult moral and political issues, this does not negate the basic point that buying slaves should not 'be equated (in terms of power) with ruling the country where they were bought' (Austen 2008: 1004). As we shall see, Europeans generally found it difficult to dictate terms easily or consistently to their Atlantic African counterparts prior to the mid-nineteenth century, so their relationships with their Africa counterparts displayed a number of different attributes to the comparatively more one-sided affairs commonly associated with the Scramble of the 1880s and beyond. These differences have frequently been overlooked due to a recurring tendency to project European dominance backwards through time. Additionally, the presumption of 500 years of constant European dominance tends to have the effect of minimizing African agency and autonomy by relegating Africans to the role of pawns and passive victims at the hands of more dynamic and sophisticated Europeans. This image can make it difficult to imagine futures and relationships that take place outside the context of Western domination, since external domination is presumed to be an historical constant, rather than a relatively recent and historically exceptional development.

Our favoured approach adheres to a now substantial literature (e.g. Northrup 2008: 55; Morgan 2009: 225).[1] In the African context, one key early example comes from Philip Curtin (1964: 8–9), who argued that:

> relations between African political authorities and Europeans trading on the Coast were generally those of equal partners in a commercial transaction … political sovereignty was seldom raised in the technical sense of European usage. There were some trading forts on shore, but the relative power of

> the European traders and the African states was roughly in balance. A trading fort did not necessarily imply a sphere of European influence over its hinterland. In most places, the forts were allowed by Africans as a mutual convenience for African and European merchants alike.

This image also resonates with the findings of other chapters in this volume, which demonstrate that the early modern rulers in places such as China, Japan, Turkey and India also interacted with European interlocutors from a position of superiority or equality.

Nothing in the proceeding remarks should be taken as diminishing or excusing the catastrophic levels of death and suffering associated with slavery within Atlantic Africa, during the Atlantic crossing, or within the Americas (Quirk 2009: 36–38, 56–58). Our goal here is not to minimize the horrors of enslavement, but instead to help understand how and why transatlantic slave trading reached such unprecedented dimensions and lasted for such a protracted historical period. Like numerous other historical atrocities and systematic abuses, transatlantic slavery was justified and sustained by a combination of overlapping and competing interests, calculations and ideologies. In order better to understand the underlying logics and associated tensions that were at work on this occasion, we need to view the early modern period as separate and distinct from later periods, rather than projecting more recent experiences and expectations backwards upon the past. To this end, the remainder of this chapter focuses on the nature of early modern Euro–African relations in the context of transatlantic slavery.

Atlantic Africa and slavery

At this juncture, it is important to clarify briefly what we mean by Atlantic Africa in the context of this chapter. As a geographical referent, Atlantic Africa refers to coastal regions and associated hinterlands that were integrated into the Atlantic world at various points from the fifteenth and sixteenth centuries onwards (Bailyn 2005; Falola and Roberts 2008). Accordingly, Atlantic Africa can be said to begin with Senegambia in the north and conclude with Benguela in southern Angola. As such, it comprises two regions more commonly known as West Africa and West Central Africa, which are usually separated from one another using Cape Lopez as a marker.

Since slavery became central to patterns of cross-cultural exchange, these boundaries were chiefly determined by which areas consistently participated in maritime slave trading. Deep-water vessels did not play a major role in Atlantic Africa prior to European incursions, so the arrival of ships capable of navigating the Atlantic Ocean initiated all kinds of political, cultural and economic adjustments (Morgan 2009: 223–24; Northrup 2008: 10–16). While direct European penetration into most parts of Atlantic Africa rarely went much beyond coastal sites, European traders had a further indirect impact upon hinterlands from which slaves, gold, ivory, hides, pepper, tropical gums, dye woods, palm oil and other goods were sourced in the interior and transported to the coast (Hopkins 1973: 87–112; Eltis and Jennings 1988; da Silva 2011: 202–11, 247–70). Hinterlands were not static, but dynamic, and were therefore subject to complex – but frequently difficult to quantify – forms of expansion and contraction (Harms 1981: 24–30; Miller 1988: 140–53; Hawthorn 2010: 61–65, 67–81; Vansina 1990: 200–18; Candido 2006; Nwokeji 2010: 9–10, 45–51, 60–64).

Based upon the evidence currently available, it is clear that Atlantic Africa was far from being a cohesive or homogenous region when seen from the perspective of its diverse inhabitants. According to historian John Thornton (1999: 15), it is possible to 'name and locate at least 150 sovereign polities' within Atlantic Africa around 1600. The incomplete nature of available information means that this figure is probably an underestimate. Some of these 'sovereign polities', such as Kongo, Oyo, or Dahomey, covered comparatively large territories and had substantial populations, but it is estimated that 'probably half of the people in Atlantic Africa lived in polities that measured around 50 kilometers across and had only a few thousand inhabitants' (Thornton 1999: 15). This profile has been reflected in discussion of differences between centralized and decentralized governance, and the attendant consequences of these differences upon patterns of slave trading (Klein 2001: 49–65; Miller 1976: 1–11, 266–79; Inikori 2003: 176–91; Nwokeji 2010: 11–19).

Unsurprisingly, relationships between various rulers and communities could take different forms at different times, with trading relationships, tributary arrangements and political alliances either alternating or overlapping with organized violence, ranging from occasional raids to protracted wars. Since direct European penetration was limited, relationships between regional neighbours within Africa were typically far more important than relations with Europeans. Instead of Europeans occupying a paramount position, the most common dynamic involved situations in which European inputs influenced the way in which relations between neighbours transpired.

This pattern has some resonance with a now-familiar distinction within IR circles regarding the difference between an international system and an international society, with European involvement in Africa during this period having been most commonly framed in terms of the latter. While it is true that cross-cultural socialization was generally fairly limited during this period (see below), the language of an international system is not necessarily ideal here, since what was primarily at issue was the development of institutions and agreements for regulating and facilitating the movement of ideas and items across cultural divides. An international system implies that both Europeans and Africans were substantially integrated into an overarching system which meant that their military and diplomatic calculations were principally defined in terms of other actors in the system. Outside Portuguese Angola, the direct military and diplomatic dimensions of Euro–African exchanges were relatively muted, as African elites were far more concerned with their regional rivals within the continent than Europeans, and Europeans were similarly unconcerned with the prospect of military invasions or alliance formation from Africa. The prototype for discussion of system versus society is the role of the Ottoman Empire within Europe during the early modern period (Göl, this volume), where the Ottoman presence was central to a cross-cultural regional balance. Since Euro–African relations were organized along very different lines, we think it is more useful to think in terms of cross-cultural institutions and arrangements, rather than international systems or societies.[2]

Both geographic and demographic considerations also played a substantial role in structuring patterns of settlement, trade, warfare and social organization. While a full consideration of these issues cannot be attempted here, two brief observations might be offered. First, on the question of geography and ecology, it is important to underscore the general point that the regional distribution of navigable rivers, rainforests, savannahs, access to coastal trading sites, weather conditions and disease environments all played an important role in shaping relationships both amongst Africans and between Africans and Europeans (Connah 2001: 108–79; Thornton 1999: 14–15). In keeping with patterns in other parts of the globe, rivers proved particularly efficient vehicles for trade, but in their absence overland travel frequently proved challenging and time consuming, though not impossible. Horses were not uncommon in parts of Atlantic Africa during the pre-colonial era, but their overall impact was complicated by disease and inhospitable terrain in many settings (Law 1980; Elbl 1991: 85–110). According to Jeffrey Herbst (2000: 35–57), these sorts of factors contributed to an environment where it proved difficult to 'broadcast' power owing to the high costs associated with expanding and maintaining territorial control over large domains.

Herbst's (2000: 15–16, 37–39) work also provides the basis for a second observation regarding the complementary role of demographic factors. Many parts of Atlantic Africa had relatively low population densities compared to much of China or Europe. This is widely held to have contributed to an environment where land tended to be abundant while labour remained scarce (although there are important exceptions to this pattern, such as parts of the Bight of Biafra). Since land was frequently abundant and agricultural production was relatively flexible (the absence of the plough meant that agriculture required little investment in specific plots), communities faced with external threats were not always obliged to defend valuable fixed infrastructure or major urban centres, and could therefore more easily opt to migrate and rebuild their lives at comparatively manageable costs. Consequently, control over territory tended to be less important than control over people in many – but by no means all – cases. Instead of primarily focusing on conquering fixed territories, many elites regularly focused their efforts upon both controlling and acquiring people.

Slavery frequently played a prominent role in these efforts. The history of slavery in what became Atlantic Africa long predated European oceanic contact in the fifteenth and sixteenth centuries, though there continue to be debates about its geographical extent and density (Nwokeji 2011). While slavery as a general institution has been reported at most times and places throughout history, scholars have also documented many variations both within and between slave systems. This was as true of Africa in general and Atlantic Africa in particular, as of many other places. It is not possible to do these variations justice here (see Quirk 2009: 23–33). For the purposes of this discussion it is sufficient to make two further observations. First, while slave systems in Atlantic Africa were often organized on quite different terms to the more familiar plantation slavery of the colonial Americas, there were nonetheless sufficient similarities in the proprietary claims associated with slave systems in Africa to facilitate extensive cross-cultural slave trading. Second, slavery in Atlantic Africa was especially valued for its capacity to incorporate – and thereby effectively exploit – significant numbers of new members into political communities which were chiefly organized in terms of social hierarchies, kinship structures and 'rights in persons'.

As a general rule, control over people rested upon complex forms of dependency and patronage, with junior members within a community providing services, support and material goods for their superiors, while their superiors in turn offered their qualified patronage and protection. Within this hierarchy, problematic or uncooperative individuals could be disciplined in various ways, including violent coercion, enslavement and sale. Concurrently, raiding and trading for slaves frequently helped political and economic elites to consolidate and expand their position within prevailing social hierarchies, and also provided an avenue for more junior members to improve their standing (Harms 1981: 31–39; Miller 1988: 40–62; Law 1991: 63–69).

One of the main ingredients of slavery in Atlantic Africa (and elsewhere) during this period was natal alienation, which refers to the status of slaves as 'kinless' outsiders, whose lack of socially recognized kinship ties placed them at the bottom of the social hierarchy. Being 'socially dead', slaves – or at least recently acquired slaves – tended to be extremely vulnerable, and were consequentially routinely exploited and abused (Miers and Kopytoff 1977; Patterson 1982; Smallwood 2007: 59–64). Many enslaved Africans were not traded to Europeans, but instead ended up in local, regional or trans-Saharan slave markets. While labour exploitation was ubiquitous, slaves were also forced into service as soldiers, sacrifices, functionaries, concubines and junior wives. In many cases, female slaves and children were especially prized, with the former being valued for both their productive and reproductive capacity, and the latter being valued as being more amenable to socialization and control (Campbell *et al.* 2007, 2009; Roberston and Klein 1988). As we shall see, European incursions had a profound impact upon patterns of enslavement both in Atlantic Africa and in the colonial Americas.

Early European incursions and the Portuguese in Angola

Despite reaching many parts of Western Africa before the discovery of the 'New World' in 1492, Europeans rarely succeeded in establishing colonial settlements on the continent until after 1850. This pattern of interaction with Atlantic Africa can be contrasted with concurrent developments in the Americas. By 1550, Spanish and Portuguese invaders had established major bridgeheads in the Americas, having conquered both the Aztec and Incan Empires. By 1650, the Atlantic coastline of the Americas was familiar to many Europeans, and the British, French and Dutch had also begun to establish colonial footholds by that date. A century later, large portions of the Americas were firmly under European rule, although native Americas continued to exercise significant amounts of agency and autonomy in some settings (Jones, this volume). This pattern in the Americas can be contrasted with settlements in Atlantic Africa in 1750, where Europeans remained confined to a series of modest forts along the Atlantic littoral and a more substantial Portuguese settlement at Luanda. As we explore further below, most settlements existed under local sufferance and their primary goal was to establish local rights to trade against European rivals, rather than to establish territorial control. By the mid-nineteenth century, peoples of European origin had mostly completed the colonization of the Americas, yet they knew relatively little about much of the interior of Africa, despite centuries of trade.

Portuguese agents were the first Europeans to make contact with local peoples along the Atlantic coast, and they represented the only European state to establish a major African colonial settlement before the early nineteenth century. Commercial motives played a central role from the outset, with gold, ivory and slaves being prominent as trade goods, yet the Portuguese were also keen to acquire new territories, recruit allies and convert Africans to Christianity (da Silva 2011: 231–70; Boogaart 1992: 369–85). Starting with the Canary Islands in the early fourteenth century, Portuguese vessels moved south along the coast, reaching Sierra Leone in 1460, colonizing the uninhabited Cape Verde islands during the 1460s, and established forts along the Gold Coast (Elmina in 1482, Axim in 1515, and Shama in 1523), sugar-producing colonies on the islands of São Tomé and Príncipe, and later a major presence in what became known as Angola. Some of these initial encounters were violent, with Portuguese sailors seeking to enslave local Africans, but these efforts provoked strong resistance, and the focus shifted from raiding to trading, with thousands of slaves being taken to Portugal prior to and even after the 'discovery' of the Americas (Thornton 2007: 138–44).

In addition, the Portuguese both sent and received a significant number of diplomatic envoys. From the 1480s onwards, a series of envoys from Atlantic Africa travelled to Portugal: 'The first, in 1484, came from the kingdom of Kongo on the lower Congo River. The king of Benin sent an embassy to Portugal in 1486. Delegations from the Jolof kingdom on the lower Senegal river came to Portugal in 1487 and 1488 … [There was a] new Kongolese embassy in 1488–90' (Northrup 2008: 5). These diplomatic exchanges had both a political and religious component, with the Portuguese making efforts to convert elites from Africa to their faith. Significantly, a number of rulers and other elites were baptized during these early encounters. These conversions were influenced by efforts to secure military and supernatural support. In the decades following initial contact, small numbers of Portuguese troops fought alongside local allies in Senegal, Benin, Sierra Leone and Kongo, but it was only in the final example that a major and sustained presence emerged, since in most cases both diplomacy and conversion had a limited long-term impact.

During this period, the rulers of Kongo are estimated to have governed 'something in the order of 100,000 [square] kilometers with as many as 350,000 people paying taxes and accepting justice from the kings at Mbanza Kongo', the capital city (Heywood and Thornton 2007: 53). Contact between Kongo and Portugal began in 1483, with the arrival of a small fleet led by Diogo Cão, which prompted a series of exchanges that saw a steady stream of envoys travel to Portugal (and later Rome), while Portuguese agents in turn established themselves in Kongo. Starting in 1491, successive rulers of Kongo were baptized into the Catholic faith, with the long reign of Afonso I Mvemba a Nzinga (1509–42) proving particularly consequential in establishing churches and schools, and also paving the way for missionaries and other settlers. The uptake of Portuguese ideas and influences tended to be most pronounced amongst elites and within urban centres, but even within elite circles local traditions often retained legitimacy and support, making it necessary to think in terms of both varying degrees and varying combinations of local and European influences.

For Linda Heywood and John Thornton (2007: 67), these exchanges resulted in the emergence of 'Atlantic Creoles', a new cultural form that was:

> best represented by the profession of Christianity but also by knowledge of European languages (and some literacy) and political ideas, adopting mixed European and African names … the adaptation of imported cloth and clothing items from both Europe and West Africa, some mixing of musical styles, and the absorption of American food groups and preparation techniques.

While there is no question that these cross-cultural exchanges resulted in the emergence of novel identities and social formations, the precise terms on which they should be conceptualized continues to be debated. One of the issues at stake is the extent to which classifying those involved as African 'Christians' unduly privileges European referents, and thereby overlooks the centrality of local forms of community and cosmology. Particularly significant in this context is the work of James Sweet (2003: 113), who maintains that:

> Christianity and indigenous Kongolese religion operated in *parallel* fashion, with the broader Central African cosmology still being the dominant religious paradigm for most Kongolese … [who were] likely using Christian symbols to represent their own deities, and they continued to worship them as they always had.

This is not to say that religious convictions were not always genuine, but it also appears that the terms on which Christian doctrines were conceptualized and operationalized owed a large debt to local models. It is likely that Europeans frequently saw more of themselves in Africans than was actually the case.

It is also clear these new cultural influences were mediated by larger political and economic considerations, both in terms of relations between the Portuguese and the Kongolese, and between Kongolese elites and their subjects and neighbours. Slave trading proved to be especially important in this context, since gold and spices found in other parts of Atlantic Africa were not locally available in Angola. Initial demand for Angolan slaves chiefly came from São Tomé, but this market was overshadowed by the emergence of a rapidly growing Brazilian market by the mid-sixteenth century. From the outset, warfare and slave trading were inextricably linked, with Portuguese soldiers fighting on the side of Kongolese armies, and sometimes taking part of their payment in slaves. Raids and wars against neighbours also accelerated – sometimes with Portuguese involvement – to help meet the growing export demand. This marked the beginning of a far-reaching transformation, as demand for slaves provided a catalyst for increasingly organized violence. Kongolese rulers at first exercised a high level of control over this trade, both in terms of regulation and revenue raising, but over the course of the sixteenth century events gradually began to escape their control, with private traders, invaders and regional rivals all intruding upon the status quo. This created a volatile environment which the Portuguese were eventually able to exploit to their advantage using a complex series of alliances and collaborations.

Portuguese agents first moved towards establishing a colonial settlement at Luanda in the 1570s, nearly a century after Diogo's initial arrival. This move can be traced to a combination of different factors. During the late 1560s, much of Kongo – including the capital – was overrun from the east by Imbangala invaders, whose forces were only finally repelled with the help of 600 Portuguese soldiers dispatched from Lisbon. This exposed the weakness of Kongo and its reliance on Portuguese support, while also revealing the potential vulnerability of Portuguese traders and settlers in the region (Birmingham 1966: 42–48). Following the invasion, the aggressively minded Portuguese King Sebastião granted a Royal Charter to Paulo Dias de Novais. In 1571, Novais was accorded extensive economic privileges in exchange for an ambitious effort to 'subjugate and conquer the Kingdom of Angola' and establish a new colony (Heywood and Thornton 2007: 82).

Drawing upon his previous experience in the region, Novais launched his colonial venture at Luanda, which was to the south of Kongo. Between 1575 and 1592 some 2,340 Portuguese arrived in Luanda, but sickness and warfare meant that only around 300 still survived in 1594 (Curto 2004: 56). As a consequence, the viability of the new colony – which was subsequently taken over by the Crown following Novais's death – was heavily reliant upon the capacity of the Portuguese to build alliances with local *sobas* and other groups such as the Imbangala (Miller 1976: 176–223).

Despite suffering defeat at the hands of the Ndongo and Matamba in 1590, the Portuguese were able to retreat, regroup and eventually secure new allies. Thus, with Luanda serving as their primary base on the continent, the Portuguese expanded both their direct territorial rule and their larger sphere of political and economic influence over the course of the sixteenth century and beyond. There were a significant number of hurdles and reversals along the way, including a disastrous invasion of Kongo in 1622–23 as well as the occupation of Luanda by the Dutch (then allied with Kongo) between 1641 and 1648, but these ultimately proved to be temporary setbacks that did not prevent the long-term consolidation of Portuguese control and influence. The process was facilitated by internal fragmentation in Kongo, Ndongo and other regional powers, and took place in the context of constant, multi-sided warfare. In time, the Portuguese were able to establish settlements and forts in many parts of the interior and at other coastal sites such as Benguela (Miller 1988: 230–31; Miller 1976: 196; da Silva 2011: 188–90). At no stage, however, were the Portuguese able to do away with their often autonomous local allies and instead exercise dominance on their own.

The political economy of Atlantic slavery introduced a series of incentives which played a key role in enabling the Portuguese to recruit allies in Africa. The slave trade could not have taken place without sustained levels of African participation/collaboration as trading partners and/or military allies. In the case of Angola, large numbers of slaves were acquired through wars and raids that were routinely initiated by the Portuguese, yet were either chiefly or exclusively fought by their African allies (Miller 1988: 196–97). Both European credit and complex combinations of local and European currencies helped to facilitate trading from the interior to the coast.

Both the terms of these exchanges and the composition of Atlantic trading networks are considered in more detail in the following section. As we shall see, there were a number of areas where cross-cultural exchanges in Angola displayed features similar to exchanges elsewhere in Atlantic Africa. The main characteristic that distinguishes Angola from other African regions is the extent to which the Portuguese penetrated the interior, rather than being reliant upon coastal enclaves or shipboard trade to sustain their activities. The extent of this penetration resulted in a distinctive form of international order which reconfigured established regional models in relation to patterns of warfare and trade, complex forms of cultural diffusion, and the emergence of a substantial long-term Euro–African community which synthesized African and European influences. It would be misleading, however, to approach this international order as an expression of Portuguese dominance. Their unusual success first in Kongo and later in Luanda was far from inevitable, but instead emerged from a contingent and opportunistic process in which the Portuguese only secured the upper hand centuries after first contact.

In order better to understand this historical trajectory, it is important to recognize the role played by: a) personal interests and collective identities in motivating behaviour; and b) the importance of short- and medium-term calculations in shaping decision making. It is only by taking into account these different perspectives – and recognizing distinctions between individual gains and collective costs – that we can begin to make sense of the characteristics and consequences of transatlantic slavery in Atlantic Africa. When elites within the region decided to establish links with the Portuguese (or other European agents), it was because they calculated that this would be to their immediate personal advantage and/or that of their families and communities. Not all of these calculations were subsequently borne out by events, but this does not mean that they were unreasonable at the time they were made. While slave raiding and trading invariably meant violence and death, for slave suppliers this represented an acceptable cost, not a decisive consideration, because the carnage associated with slavery was primarily directed towards enemies and/or outsiders from other communities. The capacity of slave suppliers to externalize (albeit sometimes only partially or temporarily) the social costs of procuring slaves to sell to Europeans was fundamental to the political economy of slaving in Africa.

In any discussion of this period, it is important to keep in mind that 'the terms "Africa" and "African" and the perception that the continent of Africa (or at least the sub-Saharan portion of it) comprises a unified cultural and/or "racial" unit are European in origin' (Sidbury 2007: 6; see also Diouf 2003: xiii–xv; Miller 2012: 79). Consequentially, the fact that enslavers and enslaved shared a common African origin was almost entirely overshadowed by the fact that they came from different communities and/or ethnic backgrounds. While slave trading undoubtedly inflicted a heavy human cost on Africa, these burdens were not distributed evenly. Socially constructed divisions between insider and outsider were central to an underlying cost-benefit calculus that encouraged slave trading. In this respect, 'Atlantic Africa', much like 'Africa' in general, most strongly emerges as a cohesive region from the perspective of European outsiders and Atlantic trade – or with the benefit of hindsight – since African participants were more likely to use more localized referents. There was not one integrated 'Atlantic African' international order, but many overlapping international orders in a region defined by similar cross-cultural understandings and mechanisms designed to regulate and regularize trade.

Cross-cultural institutions and arrangements in Atlantic Africa

In this section we are particularly concerned with: a) African agency in shaping the content and conduct of trade; and b) Euro-African institutions and arrangements that facilitated movements across cultural and political lines. While Europeans were unable to replicate the scale of the colonial settlement established by the Portuguese in Angola, they were able to establish long-term trading relationships in many parts of Atlantic Africa. As the scale of European trade grew, slave trading acquired a dominant position in many locations from the mid-seventeenth to the early nineteenth centuries. As part of this process previous trading networks associated with other commodities such as gold and ivory were gradually superseded. By the second half of the eighteenth century between 40,000 and 60,000 slaves were being embarked annually via a regional network of ports (Eltis and Richardson 2008, 2010). It was not uncommon for coastal rulers and traders to welcome ships from more than one European country. On the Loango Coast, to take but one example, Dutch, English and French slavers all plied their trade alongside each other during the eighteenth century (Sommerdyk and da Silva 2010). While not all nationalities traded on all parts of the coast, competition amongst traders nonetheless played a recurring role in relation to prices, bargaining and trade goods.

Local conventions, politics and preferences all tended to have a major influence over how trade took place. On many parts of the coast, traders were obliged to offer tributes and pay taxes to various elites, and/or to adhere to various protocols, displays of social deference, or other rituals (da Silva 2011: 198–202). The precise content of these conventions usually reflected a combination of geographic and political considerations. In the case of the former, natural features such as harbours, lagoons and anchorages regularly influenced how European vessels established contact and conducted their business. In the case of the latter, local rulers often went to significant lengths to confine European traders to designated locations and trading protocols. At some ports, such as Cape Coast (Gold Coast), Anomabu (Gold Coast) and Ouidah (Bight of Benin), there emerged small permanent settlements with European and Euro-African resident populations who brokered trade. This can be contrasted with other ports such as Cape LaHou (Windward Coast), Bonny (Bight of Biafra), Old Calabar (Bight of Biafra) and Loango, Cabinda and Malemba (West Central Africa), where trade mostly depended on ship-to-shore interaction, though temporary storehouses were sometimes erected on shore (Galenson 1986: 22–23). During ship-to-shore trade Europeans were sometimes obliged to return to their ships to sleep while conducting trade, which offers a compelling illustration of how their movements were circumscribed and regulated by their African trading partners.

Most permanent settlements operated under varying degrees of local sufferance. Some settlements involved separate fortifications, such as Cape Coast and Elmina, but these fortifications had limited capacity to project power inland. Their purpose was instead to guard against encroachment from other European powers. 'Despite a fortified presence on the coast stretching back sixty-five years, a Dutch factor stated in 1702 that "no European nation can feel safe on this coast unless the surrounding Natives are on its side' (Eltis 2000: 147). Other settlements took the form of 'forts', 'factories', or 'villages' either within or adjacent to local coastal towns. One prominent example of this pattern comes from Ouidah, where the French, English and Portuguese all established separate quarters and 'great houses' within the town (which was 3 km inland, and thus not a 'port' in the conventional sense).[3] Significantly:

> there was never any question that the European establishments were in the final analysis subject to local control, rather than representing independent centres of European Power. This was explicitly addressed in the policy of the Hueda kings of forbidding fighting amongst Europeans in the kingdom, even when their nations were at war in Europe.
>
> (Law 2004: 36)

While limited cultural exchanges (such as intermarriage and the presence of African translators) helped to facilitate trade in particular cases, long-term trade also routinely took place in the absence of shared cultural values. The European presence was, in any case, usually fairly small. In the early 1700s, the English fort at Ouidah was 'manned by only 20 white men, with 100 "gromettoes", or African slaves', within a town of perhaps 1,000 to 1,500 (rising to perhaps between 6,000 and 8,000 by the late eighteenth century, when Ouidah was ruled by Dahomey) (Law 2004: 39, 73). Despite this modest size, Ouidah nonetheless functioned as a key site through which more than 1 million slaves were embarked (Eltis and Richardson 2010: 90).

In a number of cases, slave trading was mostly conducted via trading networks rather than via local rulers, who preferred instead to focus on taxes and tribute. Over the course of the transatlantic slave trade many communities of traders expanded within coastal settlements and/or associated hinterlands. These communities most commonly served as intermediaries or middlemen between inland suppliers of slaves and African coastal traders and/or European buyers. Traders were sometimes bound together by a common identity or (fictive) kinship (notwithstanding a degree of internal diversity), with social outsiders from different backgrounds frequently being (at least initially) confined to subsidiary or supporting roles. These kinship ties could play a major role in helping to facilitate long-distance trade in situations where authority was relatively decentralized and/or localized. Communities of traders could be found at coastal sites, such as Bonny, Old Calabar, Lagos and Ouidah, and also within the interior, such as the Aro and the Duala in parts of the Bight of Biafra and the Bobangi in the Congo basin (Latham 1973; Harms 1981; Austen and Derrick 1999; Law 2004: 71–122; Lovejoy and Richardson 2004: 363–92; Mann 2007; Nwokeji 2010).

The fortunes of all these communities were closely connected to the growth of Atlantic trade, as the scale of their operations, populations and networks expanded significantly in response to demand for slaves and other goods from Europeans. Here, as elsewhere, slave trading represented a key vehicle for increasing political power and economic wealth, both personal and communal. This was not simply a question of the political economy of wartime enslavement, although this certainly remained a key issue through this period. It was also a matter of communities that pre-dated European contact undergoing long-term transformations as a consequence of a political economy that incentivized particular types of trading practices and relations.

These arrangements and institutions not only emerged within Africa, but also extended to Euro–African relationships. An important ingredient in allowing local wealth to grow was the trade credit that many African slave dealers regularly and perhaps increasingly received from European traders (Lovejoy and Richardson 1999, 2007; Law 1991: 217–19). While the specific content of these credit arrangements took different forms at different times, the most common model involved Europeans providing trade goods that would then be transported into the interior and exchanged for slaves. Once sufficient slaves had been procured, they were taken back to the coast on caravans or by boat to be traded with Europeans, with the value of the goods that had been provided being factored into the exchange. This was effectively a system of deferred payment, wherein Europeans advanced goods in the expectation of being appropriately compensated later. These transactions rarely operated on good faith alone, however, but were instead supported by mechanisms designed to ensure repayment. Sometimes credit was underwritten by the authority of a local ruler, as at Bonny in the Bight of Biafra, but the most common method of underwriting credit was to rely on pawnship as a way of providing financial collateral. Pawnship was a longstanding and widespread African mechanism for securing credit which predated the transatlantic slave trade to the Americas. While bonded, the pawn was expected to labour, but their exertions would not count towards the repayment of the original debt (Lovejoy and Falola 2003).

As adopted in trade with Europeans, pawnship typically involved individuals – often junior dependants – being placed with creditors as collateral for trade goods. When they later returned with slaves from the interior, the pawns were redeemed, but if the African traders failed to live up to their side of the bargain the pawns became slaves. Disputes over contracts involving pawns were not uncommon, and might be referred to local secret societies, such as the Ekpe (or the Leopard) Society in the Bight of Biafra and the Poro Society in Sierra Leone. Hierarchical in structure and male dominated, these also counted contract enforcement among their functions (Latham 1973: 31–51; Simmons 1956; Little 1965, 1966). Without cross-cultural mechanisms for the provision of credit and contracts supported by enforcement mechanisms, transatlantic slave trading might not have reached the level it did.

Some exchanges of goods for slaves in Africa were based on simple forms of barter between buyer and seller. In most cases, however, the complexities of trade negotiations and the varieties of commodities involved required more sophisticated means of exchange. Consequentially, the vast majority of transactions ultimately came to be negotiated in coastal currencies, such as the 'bar' (Gambia, the Windward Coast and Bonny, albeit with different values in each location), 'trade ounces' (Gold Coast and Ouidah), 'coppers' (Old Calabar), and 'pieces' or 'cloths' (Loango and Angola) (Richardson 1979: 318, 323; Johnson 1966; Northrup 1981: 157–64).

Coastal currencies provided a mutually intelligible framework for cross-cultural trade, with both Europeans and African traders contributing to new forms of standardized currencies that were a synthesis of different cultural understandings and conventions. One of the most interesting goods to be exchanged for slaves was cowry shells, which were imported via Europe into West Africa from the Maldives. Africans may have first become familiar with cowry shells through imports via East Africa or across the Sahara, but when first the Portuguese and then later the Dutch and British discovered cowries being used as currency in West Africa, they also began to import significant quantities. These cowries had previously entered Europe as imports from the Indian Ocean before being purchased by slave traders, most often in Amsterdam, the Maldives and India. Cowries were first transported to Europe and then re-exported to an expanding 'cowry zone' in Africa that centred primarily on ports at the Gold Coast and the Bight of Benin. Within this zone, they featured prominently amongst trade goods exchanged for slaves, and thereafter were transported inland in large quantities, where they served as currency (Hogendorn and Johnson 1986).

In addition to adhering to local institutions and arrangements, European traders also had to satisfy local preferences for imported trade goods. It has sometimes been argued that Europeans duped Africans by offering poor-quality goods, particularly in the case of firearms. While inferior goods and duplicitous trading were not unknown, consumers in Africa were actually fairly discerning, and their preferences had to be accommodated (Law 1994: 50–64; Thornton 1998: 7, 123–25; Eltis 2000: 137–63). This was reflected in the 'bundles' or 'assortments' of goods that Europeans had to supply if they wanted to compete in buying slaves (Richardson 1979: 66–70; Lovejoy 2000: 106–11). By the height of the transatlantic slave trade during the eighteenth century, the terms of trade in Atlantic Africa had generally moved against Europeans and in favour of African slave suppliers, yet Europeans continued to participate despite the increased costs due to the value of slaves in the Americas (Gemery *et al.* 1990: 157–77; Richardson 1991: 21–56).

In addition to currencies such as cowries, the principal goods that were exchanged for slaves were textiles, alcohol, weapons, and other metal wares, such as bar iron and copper basins (Alpern 1995). Bundles of trade goods varied significantly not only between places but also over time. The textiles used in trade bundles comprised a great variety of fabrics, and included East India textiles re-exported to Africa via Europe. Indian printed and dyed cotton textiles were in demand throughout much of Atlantic Africa, but came to be prominent in Loango and Angola, with (re)imports from India securing a significant market share by the second half of the eighteenth century. Indian textiles were only one of a number of trade goods that originated from outside Europe. In the case of alcohol, large quantities were also exported from Brazil, with around 1,342,500 litres – the vast majority being cachaça – imported to Luanda and Benguela between 1782 and 1830 (Curto 2004: 192). Imported trade goods were widely disseminated in African markets, reflecting a political economy of slavery that saw goods being dispersed amongst trading networks and communities. Atlantic imports occasionally displaced local products, but the most common scenario involved Europeans supplying similar types of goods to those already available within local markets and interior trading networks, so their impact was mostly a question of increasing consumer choice in relation to volume and variety (see Martin 1972: 158–74; Northrup 1981: 164–76; Latham 1973: 24–25; Lovejoy 2000: 276–89).

Both the conduct and content of patterns of cross-cultural slave trading provide further evidence that trade was more often negotiated than imposed. If traders and elites in Africa had not benefited, or at least not expected to benefit, then long-term trade simply would not have taken place. Moreover, when trade did occur, it was conducted on the basis of significant African input. As we have seen, Europeans rarely penetrated the interior, but instead found that their movements were limited and regulated, and as a consequence their direct cultural influence was also relatively muted (although the impact of goods exchanged for slaves may be another matter). Although there were some occasions and contexts, particularly in maritime settings, where Europeans enjoyed advantages, the general picture that emerges is one where African agency consistently played a central role. If Europeans had enjoyed a decisive military or political advantage over their African counterparts *on land*, then it is likely that things would have played out differently. Instead of carrying slaves to the Americas, Europeans might have instead moved to create colonies on the African mainland similar to parallel ventures in the Americas (Eltis 2000: 139).

Concluding remarks

Europeans have inflicted many horrific abuses upon Africans and peoples of African descent. These abuses cannot be reduced to isolated incidents or 'aberrant' policies or cases, but instead need to be understood in terms of far-reaching patterns of systematic exploitation and discrimination, severe abuse, and high levels of premature death. Furthermore, it is important to keep in mind the additional consequences of European policies and practices based upon ignorance and arrogance, such as the imposition of alien borders and the (re-)invention of 'ethnic' categories under colonial rule. The legacies of these abuses remain with us in numerous ways. By any reasonable standard, Europeans have a great deal for which to apologize. This does not mean, however, that *all* historical relations between Europeans and Africans can be either defined or explained in terms of the dominance of the former over the latter. As we have seen, cross-cultural relations between Europeans and Africans during the early modern period – at least as far as the slave trade in the Atlantic world is concerned – were generally organized on quite different terms to those of the more recent colonial and post-colonial periods.

The one major colony that did emerge – Angola – was built upon high levels of local collaboration and cooperation, an unusual window of opportunity and long-term consolidation. Unlike the numerous colonies that were rapidly established elsewhere in Africa during the late nineteenth century, the Portuguese took over a century to establish a colonial foothold once contact had been established, and their capacity to project their power far beyond the coast subsequently proved to be a long-term, piecemeal process that remained contingent on local alliances and collaborations. Moreover, Angola stands out as a notable exception within regional patterns, which saw Europeans otherwise largely confined to coastal enclaves.

When we look more broadly, it becomes clear that the conduct and content of trade were based upon complex negotiations and adaptations, with Europeans often finding it necessary to adhere to local conventions and to supply goods based on local preferences. Throughout this period, slave trading represented a valuable means of economic and political advancement for many traders and rulers in Atlantic Africa, Europe and the Americas. Their overlapping interests contributed to the emergence of cross-cultural institutions and arrangements that imperfectly regulated and regularized Atlantic slave trading through the provision of credit, the enforcement of contracts, the expansion of African trading communities, the standardization of cross-cultural currencies and units of exchange, and the establishment of protocols governing when and how trade would occur. While Europeans contributed to the overlapping international orders that were created and sustained by these arrangements, both their origin and operation frequently owed a debt to the adaptation of pre-existing models in Africa. It is important to recognize, moreover, that these institutional arrangements were not unique or exceptional, but can instead be better understood as a representative example of a larger historical pattern of various kinds of long-term international order(s) operating in the absence of a shared culture or civilization. These may not amount to international societies, or even systems, but they nonetheless constitute a neglected element of cross-cultural exchange prior to European expansion in the nineteenth century (Quirk and Richardson 2009).

Notes

1 This does not mean, of course, that there is universal consensus. For alternative perspectives, see Kaba 2001; Blackburn 2012: 85–87. Moreover, the long-term consequences of transatlantic slavery for Atlantic Africa have generated considerable discussion, with opinion being divided along a spectrum that runs from minor to major (Benjamin 2009: 366–71).

2 Europeans were also not the first external interlopers to encroach upon Western Africa. Various incursions and influences – including the southward spread of Islam from North Africa into Senegambia and other parts of West Africa – long predated European contact via the Atlantic and continued to strengthen as transatlantic slavery grew. See Austen 2010; Lydon 2009: 49–106.

3 On the definition of ports, see Eltis *et al.* 1999: 17–24.

8 Conclusion

Eurocentrism, world history, meta-narratives and the meeting of international societies

Richard Little

The overall aim of this book has been to reassess conventional thinking in the study of international relations about how contemporary international society and world order has come into existence. It builds on an initial premise that much of the thinking about world order and international society in the discipline is inherently Eurocentric and, as a consequence, it is argued, any assessment of how the prevailing world order came into being, or how contemporary international society will evolve in the future explicitly or implicitly rest on a whole raft of ideas that can be shown to be profoundly distorted or mistaken about many of the most fundamental features of international relations. On the face of it, this may appear to be an extraordinarily bold, indeed rather rash, claim because the contributors are effectively challenging the orientation and theoretical posture of most thinkers in the field. It is important to acknowledge, therefore, that this book is doing nothing more radical than participate in a very broad movement that is emerging across the social sciences. In other words, there is now a very widespread and growing recognition that Eurocentrism has been endemic in all the social sciences and there are, as a result, very significant debates across these social sciences about the veracity and consequences of this claim. It obviously follows that it would be very strange indeed if the accusation of Eurocentrism could not be made in the field of International Relations. Nevertheless, it is, perhaps, more than a little odd that more has not been made of this factor in the field given the importance that is attached to the issue in other social sciences.[1]

If the charge of Eurocentrism is taken seriously, however, what has happened in other social sciences is that there has been a rapid upsurge of interest in world history. It is widely acknowledged that to escape Eurocentrism it is essential to expand the spatial and temporal reach of the social sciences. In doing so, however, disciplinary boundaries are almost inevitably breached. Moreover, to develop an adequate understanding of even a relatively recent development, it becomes necessary to reach back into the past. One of the best illustrations of this development is the response to the European miracle thesis (Jones 1981). According to this longstanding thesis, still often regarded as uncontroversial, the emergence and then dominance of Europe in the global arena is explained by reference to a unique set of social and environmental circumstances that were peculiar to the European continent. The roots of this thesis extend back into the nineteenth century (and even earlier), when the social sciences began to emerge as formal and distinct disciplines in the Western world. Unsurprisingly, given the timing, there was a preoccupation with the concurrent and seemingly inexorable rise of Europe and the USA at that time. As a consequence, many of the most important and influential thinkers from the nineteenth and early twentieth centuries, such as Karl Marx and Max Weber, developed very significant and substantial accounts of the emergence and evolution of modern Europe. These accounts remain deeply embedded in the way that many social scientists continue to think about the world. However, just as the rise of Europe had a powerful effect on social scientists in the last two centuries, so the putative decline of the West is having an equally important impact on the social sciences in the twenty-first century. It is certainly one of the factors that has encouraged social scientists to challenge or reject the long-established Eurocentric perspective on global development.

Goody (2010), a social anthropologist, for example, has argued for many years that the image of Europe, rising above the rest of the globe, is an illusion precipitated by the preoccupation with, and the privileging of, European developments. By taking a world historical perspective that embraces what happened across Eurasia back in time to the Bronze Age, a quite different image begins to emerge. What we then observe is a very long period of cross-cultural exchange extending over Eurasia among a number of discrete developing areas, with Europe entering the fray very late in the day. Using this perspective, it becomes very obvious that other areas have risen to prominence in the past and will no doubt do so again in the future. Because of the persistent interaction amongst these distinct centres of development, it is more appropriate to identify a process of alternation, with one centre giving way to another, rather than to see the currently dominant centre as in some way unique. According to Goody (2010: 56), the potential to identify the continuity that has existed since the Bronze Age across Eurasia from East to West, has been 'disastrously fractured by a concentration on the European experience alone'.

Moving to a world historical perspective, however, is not necessarily sufficient to escape from Eurocentrism. McNeill (1991) frankly admitted as much when he reassessed *The Rise of the West* (1963), 25 years after his groundbreaking world history was first written. The bias is apparent in the title, which clearly reveals that McNeill's account intends to privilege one particular area of the globe, as well as some of the terminology used to launch his ambitious framework. This framework embraces four distinct civilizations that are contrasted with the nomadic 'barbarians' that occupied much of the territory separating these civilizations. The terminology, of course, does not even allow for the possibility that there could be a distinctive nomadic civilization, but this line of argument should not be used to diminish McNeill's very real achievement. As Hodgson (1993: 92), another distinguished world historian insisted *The Rise of the West* was 'the first genuine world history ever written'. Nevertheless, McNeill (1991: xvii) subsequently acknowledged that he had inadvertently looked at history 'from the point of view of the winners', and that he had been influenced unconsciously by the 'imperial mood' that prevailed in the USA after the Second World War. Later, McNeill (1993: xxiv) revised his original thesis and argued that far from the separate civilizations being insulated from each other by 'barbarian zones', they were, in practice, linked together by 'communication nets' and as he took this factor into account he began to think in terms of a 'world system' rather than separate civilizations. Indeed, McNeill (1998) came to insist that much more attention needs to be given to trans-civilization links.

What a world historical perspective can promote, therefore, is a framework that opens the way to the establishment of a grand or meta-narrative. It can be argued that the distinctively different grand narratives developed by both Marx and Weber derived from their world historical perspective. It is, of course, also possible to agree with Linklater (2009: 3–4) that these meta-narratives were 'premature' because they appeared in an era when knowledge of the world beyond Europe was still very limited. However, according to Linklater, there is now 'a consensus' across the social sciences that regards any attempt to develop a meta-narrative as profoundly regressive. Indeed, so much so, that he asserts that many social scientists now consider that they have 'a special responsibility to subvert any attempt to return to progressivist perspectives that had such damaging consequences for non-European peoples in the age of the European empires'. Drawing on Elias (1978, 1982), Linklater also insists that such a position has the effect of negating what can be seen as the essential task of studying 'long-term processes of development stretching over millennia that proceeded without any commitments to the notion of progress in history'. There can be little doubt that the resurgence of interest in world history in recent years is at least partially in reaction to the post-modern suspicion, as exemplified in the work of Lyotard (1984), of any kind of meta-narrative.

It is against this background that the current project needs to be assessed. In essence, the various authors can be seen to be responding to what they take to be the inadequacies of the grand narrative in international relations that dominates thinking in international relations and is, arguably, most closely associated with the English School. According to this grand narrative, to understand the essential features of contemporary international relations, it is necessary to focus on the evolution of international relations in Europe. It is asserted that an international society of states emerged in Europe across the course of the second millennium. For many years it was widely assumed that this society came into existence over a long period of time, although it only fully crystallized in the aftermath of the Treaty of Westphalia in 1648. More recently, the precise point in time when it can be accepted that the modern international society emerged has become increasingly contested. Some theorists push the date back in time, earlier than 1648, while others have pushed the date beyond 1648, to as late as the start of the nineteenth century.

However, there is still across these contested views an assumption that the contemporary international society finds its origins within a set of processes that were generated inside Europe. There is an important gloss on this position, at least amongst some English School scholars, who also acknowledge that the European international society existed within the context of a much more extensive international system, which eventually became worldwide before the eventual establishment of a global international society. It is argued, in other words, that states that operated in the global international system but that lay beyond the boundaries of the European international society were eventually brought into the European society which by the end of this process formed a global international society. One of the earliest entrants was, for example, Russia, although there are debates about when the entry took place or the extent to which it ever became a fully fledged member of the European international society (Neumann 2011). In any event, the key feature of the meta-narrative is that states lying outside of Europe were eventually allowed into the European club until, by the second half of the twentieth century, the European international society extended across the globe. The framework of this meta-narrative, therefore, extends over a millennium and it creates the potential for researchers working within a limited timeframe to relate their research to a much bigger scenario.

Given that this meta-narrative has been fundamentally critiqued in this book, it is important to note that the first generation of English School theorists did not consider that the meta-narrative was in any way distinctive to English School thinking. On the contrary, Bull (1984c: 123) views it instead as the 'standard European view' and, more important, he insists that it manifests obvious 'absurdities' such as the idea that states like China, Egypt, or Persia, 'which existed thousands of years before states came into being in Europe, achieved rights to full independence only when they came to pass a test devised by nineteenth-century Europeans'. Unsurprisingly, he acknowledges that it is now 'common to question this account'. On the face of it, therefore, there are some grounds for thinking that Bull was at the very least moving towards an acceptance of the critique of the 'standard account' as unacceptably Eurocentric. Elsewhere, however, Bull and Watson (1984b: 2) indicate that it is their aim to examine how Europe 'dominated, and in so doing, unified the world'; moreover, they go on to insist that 'it is not our perspective, but the historical record itself that can be called Eurocentric'. It does seem, therefore, that whatever its 'absurdities' Bull and Watson continue, in essence, to subscribe to the standard account of how the global international society came into existence.

What I want to do in the first section of this chapter, however, is to challenge this assessment and suggest that a close reading of all the contributions in Bull and Watson's edited volume *The Expansion of the International Society* generates a much more nuanced and complex meta-narrative than has generally been recognized and certainly one that is more in tune with the tenor of the chapters in this book than might at first sight seem likely. In the second section I will then assess to what extent the contributors to this book either challenge or reinforce the assessments in Bull and Watson (1984a) of how the contemporary global international society came into being.

Reassessing *The Expansion of the International Society*

Hedley Bull and Adam Watson were both chairs of the British Committee on the Theory of International Politics, a self-selected group of academics and practitioners who began meeting at the end of the 1950s with the aim of developing a deeper understanding of international relations. More recently, the committee has been associated with the establishment of the English School (Linklater and Suganami 2006). From the start, the committee operated from a world historical perspective and its members were interested in developing an approach that would allow them to compare the international societies that have emerged across the course of world history. The desire to turn this project into a book never materialized and instead the committee decided to narrow its focus and produce an edited book that would trace the evolution of the contemporary global international society. Dunne (1998: xiv) sees the resulting book as the 'culmination' of the committee's work and Vigezzi (2005: 86) regards it as the committee's 'most organic and coherent achievement'.[2] It is a substantial book, running to over 400 pages, with 24 contributors. Many of these contributors are not in any way directly linked to the English School and they certainly do not speak with a single voice. Nevertheless, it is possible to extract from this text a very distinctive framework that generates a complex but coherent meta-narrative that is distinctively different from the Eurocentric 'standard account' identified by Bull.

The Expansion of International Society is divided into four sections. The first explores 'European international society and the outside world'; the second examines 'The entry of non-European states into the international society'; the third looks at 'The challenge to Western Dominance'; and the fourth investigates 'The New International Society'. At first sight, it seems unlikely, given this overarching structure and the specific focus on Europe and the West, that it will be possible to break free from what looks on the surface to be a Eurocentric strait-jacket. However, a closer examination quickly demonstrates that there is more room for manoeuvre than is initially apparent. In the introduction, for example, Bull and Watson (1984b: 8) identify that an underlying issue for the book is the competing views of admission into the international society, with the European great powers, on the one hand, viewing admission in terms of entry into 'an exclusive club with rigorous qualifications for membership', as opposed to the view taken later by many Third World states that it was not a case of their admission but of 're-admission to a general international society of states and peoples whose independence had been wrongfully denied'.

In other words, Bull and Watson were fully aware of the argument that the European international society had at some point distanced itself from a broader international society of which the European states had previously been members. A comprehensive reading of their text indicates that they broadly support this assessment of how contemporary international society came into existence – and why Bull repeatedly makes reference to the 'absurdities' of the 'standard' view propagated by the European Great Powers themselves. Indeed, Bull and Watson (1984b: 1) make it very clear from the start that the rules and institutions of the contemporary international society 'have been shaped by North and South Americans of European stock or assimilation and also by Asians, Africans, and Oceanians, as well as by the European powers in their period of dominance'. Of course, the general assumption is that the Europeans laid down the foundations of the society and then others added to and sometimes substantially modified these foundations later. Without doubt, this is how Bull and Watson present the story at various points, but this generalized account does not fully accord with the more detailed narrative laid down in the text.

Although the origins of the contemporary international society are only sketched in very lightly by Bull and Watson, they are crystal clear that it is necessary to start the narrative at a point in time before the European international society had come into existence. At that juncture, there existed a number of regional international societies. The most important systems are identified as Latin Christendom, the Arab-Islamic system, the Indian subcontinent and its eastern extensions, the Mongol-Tartars on the Eurasian steppes, and China. All these international systems emerged, it is argued, on the basis of elaborate and long-standing civilizations. Elsewhere, in the Americas, sub-Saharan Africa and Australasia, there were less developed cultures and a geographical awareness that did not extend very far. From this perspective, then, only Eurasia was home to a number of separate and distinctive regional international systems.

The societies were also very enduring. By the end of eighteenth century, although Latin Christendom had been transformed into Europe and the Mongol-Tartars could no longer maintain their own independent international system, Gillard (1984: 87) argues that it is still possible to identify four main systems of Eurasian states: Chinese, European, Islamic and Indian. He acknowledges that the three Asian systems need to be distinguished from the European system, but he insists that they share two main characteristics: first, the members of the system 'were primarily concerned with one another and saw outside threats mainly in terms of how they could affect the internal balance of power; and second, despite the fact that their 'mutual hostility was often greater than their hostility to outside powers', most of the members of the system had a sense of being 'part of a common civilization superior to that of the other systems'. Nevertheless, by the end of the following century, all three Asian systems had collapsed and the member states had been co-opted into the emerging global international society.

From this perspective, the nineteenth century becomes a very crucial phase in the meta-narrative of how the contemporary international society emerged (cf. Buzan and Lawson 2012a), and it is certainly the case that the Bull and Watson text pays most attention to the later phases in the formation of the global international society. It is still possible, however, to identify a much longer narrative. Prior to the nineteenth century, as indicated, there were a number of regional international systems co-existing in Eurasia, although Bull and Watson are very clear that Europe needs to be distinguished from the other systems, not only because it was a very late developer, but also because it evolved along a very distinctive track. The other Eurasian international systems in the early phase are all identified, using Wight's terminology, as suzerain state systems.[3] According to Wight (1977: 23), a suzerain state is made up of a group of states that are in more or less permanent contact, 'but one among them asserts unique claims which the other formally or tacitly accept'. Throughout Europe's history as a regional international system, there were recurrent attempts by various states to establish suzerain status, but none was ever successful. This phenomenon is traditionally attributed to the emergence of the balance of power as a distinctive international institution, but it has never been convincingly explained why such an institution only successfully emerged in Europe.[4] Despite the fact that this is a crucial element in the narrative, it is not addressed in Bull and Watson. Instead, the focus is primarily on Europe's relations with the rest of the world.

Bull and Watson (1984b: 6) are very clear that to develop a coherent narrative it is necessary to adopt a global perspective from the start because they insist that the European international society did not 'evolve its own rules and institutions and then export them to the rest of the world'. Instead, the European international society is seen to be the product of two 'simultaneous processes, which influenced and affected each other'. One process involved developments within Europe itself and the other involved Europe's interactions with actors outside of Europe. So, for example, Gong (1984b: 172) notes that as relations between China and Europe intensified in the nineteenth century, there was a clash between 'fundamentally irreconcilable standards of "civilization"', but because both sets of standards were dynamic, the Sino–European interaction in the late nineteenth century and the early twentieth century, 'changed both the Chinese and European standards of "civilization"'. Gong argues, for instance, that the Chinese standard of civilization, especially in the areas of trade, diplomacy and legal relations, began to converge with the European standard of civilization, while, on the other hand, the European standard of civilization, which had in the past only been 'codified in treaties with non-European countries', now began to emerge 'in increasingly explicit terms'. As Brownlie (1984: 358) also notes, prior to the nineteenth century the Europeans had not yet 'clearly formulated a test of "civilization"'. However, overall, it has to be said that there is relatively little detailed discussion in Bull and Watson of how the process of internal development within the European international society meshed with the external interaction process with the outside world. It has been left to subsequent analysts to look at the way these two processes mesh in more detail. For example, Anghie (2004) has looked very closely at the process of interaction to show how the development of the contemporary concept of sovereignty based on recognition was very much the product of nineteenth-century European colonialism (see also Branch 2012). Nevertheless, it remains the case that the basic idea that the development of the European international society was significantly affected by its external interactions is very much foreshadowed in Bull and Watson.

By the same token, however, it is also true that the overarching framework used in Bull and Watson to link Europe with the global arena is again only very lightly sketched. Here, the starting point is that Europe begins the process of 'expansion' some considerable time before the idea of a European international society comes into existence. They argue that the development started in the late fifteenth century, although a case can be made that the process began at the start of that century, in 1415, when the Portuguese seized the Moroccan city of Ceuta. However, as Bartlett (1994) shows, Latin Christendom virtually doubled in size between 930 and 1350 as the consequence of constant colonization and so the European Christians who sailed to Asia, Africa and the Americas in later centuries had already had a great deal of experience in colonization and cultural conquest. From this perspective, overseas colonization was simply an extension of the internal colonization that had been going on for centuries. Watson (1984a: 13–14) pursues this line of argument, noting how Christendom expanded in three directions: first, outwards from the north and into the non-Christian areas south and east of the Baltic; second, to areas south and east of the Mediterranean that were occupied by non-Latin Christians; and finally to the south-west and areas like Spain and Sicily that had been taken over by the Muslims. However, Watson also insists that this third area of expansion proved crucial because 'it did not stop at the water's edge, but continued overseas to become the great maritime expansion'.

Yet, paradoxically, colonization is not an issue that is treated in any depth in Bull and Watson. Nor is there any attempt to explore the role that the colonies played in the development of the European international society. There is no acknowledgement that from the very beginning, the dominant units in the nascent European international society were empires, engaged in a process of internal colonization. As a consequence, there is no discussion of how the formation of empires played a crucial role in the transformation of the hierarchically structured Christendom through to the anarchically structured international society (Larkins 2010). This oversight, however, is not particularly surprising because it is simply a reflection of the hegemonic dominance of the 'Westphalian myth' that prevailed at that time, not only in the study of international relations, but across the social sciences and humanities. As Armitage (2000: 14) notes, 'the rise of nationalist historiography in the nineteenth century had placed the history of the nation-state at the centre of European historical enquiry and distinguished the state from the territorial empires that preceded it, and in turn from the extra-European empires strung across the globe'. Yet there is a vital link between empires and states that has simply not been observed in International Relations. By contrast, more recently, Armitage (2000: 15) argues that empires 'gave birth to states and states stood at the heart of empires. Accordingly the most precocious nation-states of early-modern Europe were the great empire-states: the Spanish monarchy, Portugal, the Dutch Republic, France and England (later Britain).' Yet, because nation-states and empires are conventionally treated as opposing political structures, the role of empires in the development of the European international society of states has been either ignored or left perennially ambiguous, in truth, in most fields of study. Even so, the failure in Bull and Watson to interrogate more closely the relationship between states and empires remains a little odd, because Heeren's *History of the Political System of Europe and its Colonies*, written at the start of the nineteenth century, places the colonies at the heart of the story and this was a book that both Bull and Watson greatly admired (Little 2008).

Although there is a failure to interrogate the role of empires in the evolution of the European international society in Bull and Watson, colonization is not, of course, ignored. On the contrary, it is a significant thread running through the entire narrative. Donelan (1984: 79–81) notes the importance of the papal bulls issued from the middle of the fifteenth century, to legitimize and support the Portuguese advance down the west coast of Africa, through to the bull *Inter Caetera* issued in 1493 that granted Spain the right to conquer the territory discovered by Columbus the previous year. The precise reason for issuing this bull is disputed but Donelan argues that probably its main purpose was to establish a line of demarcation that would help to keep the peace between Portugal and Spain. According to Watson (1984a: 17), the decision was an extension of 'medieval practice of papally authorized crusades by which the re-conquest of Iberia had been greatly assisted'. This papal act indicates, moreover, that there was still a presupposition that Christendom was hierarchically structured at this time. However, the fact that the details of the demarcation were worked out by Spain and Portugal through bilateral diplomacy, and were ratified in the Treaty of Tordesillas in 1494, also indicate that the fundamental structure of Christendom was beginning to change and give way to what was only identified from the nineteenth century as a society of states (Anghie 2004: Chapter 2). The treaty, however, did not presage a peaceful pursuit of colonies in the subsequent centuries, although Watson (1984a: 30–31) does note that the colonization that took place at the end of the nineteenth century did largely proceed on a consensual basis and that this 'consensus achieved by the Europeans over the partition of Africa takes us back to the orderly spheres of expansion agreed between Spain and Portugal, with the Concert of Europe playing the part played by the Pope'. He also observes that at this time, the propagation of Christian civilization acquired a renewed prominence', while simultaneously 'Exploitation of non-Europeans was intensified' – a further intimation that the nineteenth century was different from previous centuries.

From the sixteenth century onwards, it is argued in Bull and Watson (1984a), the Europeans acquired increasing control over the oceans and seas around the globe, but until the eighteenth century they lacked the ability to penetrate the landmasses in Africa, Eurasia or the Americas (apart from Mexico and Peru). Instead they operated largely on the periphery of all these continents. Howard (1984: 34) argues, for example, that the Europeans established 'entrepôts throughout the Indian Ocean and the China seas and, a little later, small footholds of settlement in North America'. While naval guns enabled an offshore European warship 'to defend its anchorage against land as well as sea attack', they did not allow the Europeans to project their power inland. Nevertheless, around the world, the Europeans 'were accepted by the indigenous communities on a basis of equality as useful trading partners'. Yet there were fundamental differences in the ways that the Europeans interacted with the indigenous populations in the Americas as opposed to Africa and Asia. Watson (1984a: 18) asserts that from the start, the Americas were incorporated into the European system of administration and government in exactly the same way as the Iberian Peninsula and the eastern Baltic lands had been. As a consequence, the Americas 'became an extension of Christendom', but the relationship between the Europeans and the Africans and Asians was very different. The difference is encapsulated by Watson's (1984a: 22) assessment that the:

> ... activities of competing European traders strengthened Muslim and Hindu states round the Indian Ocean economically and administratively. Informed Muslims surveying the scene around the year 1700 would see that the long Portuguese crusade against Islam in the Indian Ocean had failed, and that Islam had continued its military and religious expansion to bring almost all that great area under Muslim influence.

Until the nineteenth century, therefore, despite the growing ability of the Europeans to regulate the oceans and seas, they had no alternative but to establish relations with Africans and Asians on terms set by the inhabitants of these continents. In their introduction, Bull and Watson (1984b: 5) argue, however, that from 1500 and for the next 300 years, as relations between the Europeans and Asians persisted and grew, it is possible to identify the emergence of a 'a loose Eurasian system or quasi-system' within which the European states 'sought to deal with Asian states on the basis of moral and legal equality'. Often, of course, this was not possible. Watson (1984a: 21–22) notes, for example, that with the establishment of the Mogul Empire in the middle of the sixteenth century, the Europeans had to find a way of operating within this suzerain system and they did so, 'in the same way as the local Indian rulers, by accepting subordinate positions within the government of the empire'. The Europeans, therefore, were willing to establish formal agreements with rulers across Eurasia and, in doing so, acknowledged their sovereignty. As Brownlie (1984: 361) notes, from an international law perspective, 'there were no regional or cultural limitations on the recognition of personality in international relations' throughout this period, until the middle of the nineteenth century. It follows that the existence or sovereignty of any state 'did not depend on the recognition of other powers'.

It can be argued, therefore, that there was a nascent global international society that was slowly but steadily emerging across this period. At the start of the period, there is no doubt that the Muslims provided the main impetus behind this development because, as Watson (1984a: 17–18) notes, they were the heirs of previously highly developed civilizations, and the Turks, in particular, were 'in active intercourse with a wider range of foreign people than any other society of that time'. However, there is no doubt that because of their growing maritime supremacy from 1500 onwards, the Europeans began to play an increasingly important role in expanding the range of international contacts that were being established. Watson (1984a: 25) is equally clear that although the Europeans wished to apply their rules and institutions to others, they were willing to deal with them on their terms, but he goes on to suggest that whether we are talking of the Ottomans, Persians, Moguls or Chinese, their practices 'for all their individual peculiarities, bear a family resemblance to the arrangements that have developed at various times and places to regulate intermittent relations between different civilizations'.

During the course of the nineteenth century, however, a very dramatic transformation in the fundamental features of international relations is seen in Bull and Watson (1984a) to take place. It is argued that the nature of international economics undergoes a seismic shift as the result of technological advances that precipitated an agricultural and industrial revolution and in tandem with these developments comes an extraordinary expansion in military power. Because the USA and some of the European countries are well positioned to take the maximum advantage of these developments it was inevitable that the nineteenth century would be a period of spectacular change. However, the impact of these developments was ratcheted up because they were accompanied by some equally remarkable changes in the self-image of the Europeans and Americans. It was this factor more than anything else that transformed the nature of the international society.

The starting point for O'Brien (1984) in his discussion of the economic dimension of the expansion story is that to describe the mercantile era (1492–1789) in terms of an international economy is profoundly mistaken. He insists that the volume of transoceanic trade as a proportion of overall trade during this period was tiny and had very little impact on regional economies around the world. So his analysis represents a fundamental attack on the work of theorists like Wallerstein (1974), who assumes that we can talk about a world economy from 1500 onwards. O'Brien argues, to the contrary, that if we identify an international economy in terms of the free movement of goods, money and labour, then even at the time he was writing, it was quite clear that we were still a very long way off from having a global economy. His central point, therefore, is that the take-off of the European economy was essentially unrelated to the existence of transoceanic trade. By the same token, capital flows were 'a miniscule percentage of world capital formation' (O'Brien 1984: 48), but what we do find in the second quarter of the nineteenth century are pronounced and widespread falls in freight rates and this did 'begin to provide really strong stimulus to international trade', although '[q]uantum and qualitative leaps forward in international economic relations' did not occur, argues O'Brien (1984: 48–50), until the second half of the nineteenth century.

Howard (1984: 36–39) traces the growing military superiority of the Europeans back to the start of the eighteenth century. He argues that the growing European dominance at that time was as much cultural as military, because it required well-organized tax systems and efficient bureaucracy as well as highly developed arsenals. The level of social organization necessary to compete effectively with the Europeans did not exist in the societies with which the Europeans were coming into contact at that time. The Russians were the first of these societies to recognize that the 'only way to beat the Europeans was to imitate them, and the necessary military reforms would not be forthcoming without a huge social upheaval'. On the basis of these reforms, the Russians in the mid-nineteenth century were able to take control of Turkestan and Central Asia, and Howard argues that accounts of these Russian campaigns 'repeat almost precisely those of the British campaigns in India'.

Although the emergence of highly trained professional troops during the course of the eighteenth century was an essential ingredient of this expansion, it was the technology and scientific developments that came on stream that made it possible to open up the interiors of all the continental landmasses of the world to European and American military forces at this time. Howard focuses, first, on the development of steam power that made it possible for the Europeans to penetrate the interior of Africa up the major African rivers, but it was also possible for the Europeans to penetrate China in the same way. Where there were no available rivers, the 'speed of rail construction was astonishing' (Howard 1984: 39). Alongside these technological advances there was also the 'introduction of the quick-firing, long-range firearms made possible by the development of high explosive' (Howard 1984: 38). For Howard (1984: 38), perhaps the most crucial development was the growth in medical knowledge, because in the past disease was 'the most important barrier to European expansion and settlement'. Advances in medical knowledge made it possible for the Europeans to survive in areas that had in the past proven too inhospitable. Howard goes on to argue, however, that over and beyond this factor, it was also the case that 'European medical techniques were to do perhaps more than anything else to reconcile colonized populations to their new masters'.

This new wave of colonization in the nineteenth century, it is demonstrated in Bull and Watson (1984a), was only one aspect of a very much more substantial shift in the evolution of a global international society. At the heart of this shift was a transformation in the way that the Europeans conceived of international society and themselves. At the start of the nineteenth century, the Europeans still acknowledged that Europe operated in a global arena where groups of states operated according to their own distinctive norms and institutions. As a consequence, Bull (1984b: 111) insists, for example, that prior to the partition of Africa, 'European states displayed a considerable disposition to acknowledge that African political communities had rights of independent states' and that there is no reason to doubt 'the genuineness of treaties concluded between European states and African political communities'. By the same token, Brownlie (1984: 360) identifies a large number of extra-European political entities across Asia and Africa with which Britain signed treaties between 1750 and 1850. He asserts that with each treaty it 'must be presumed to involve an acceptance of the capacity of the other Party to enter into such an agreement with Great Britain'. However, during the course of the nineteenth century, this orientation that had prevailed for several hundred years began to shift. According to Brownlie (1984: 361–62), the change was driven by European and American international lawyers. By the middle of the nineteenth century it was agreed that state personality was determined by recognition 'and recognition was not dependent upon any objective legal criteria'. The fact of recognition was seen to dispense with the need for any formal definition of statehood because recognition, 'as the political stamp of approval, appeared to take care of the problem of definition'. With this development, statehood 'became a more important concept' and, as a consequence, a large number of political entities that had been treated as sovereign in the past were not now considered eligible to acquire statehood.

The key move that justified this change in orientation is found in the increasing reference to 'modern civilized states' by nineteenth-century international lawyers, but in fact this was a rather anodyne justification. Brownlie (1984: 362) is quite clear that this change, in practice, 'interacted with an increase in European cultural chauvinism and racial theories'. Vincent (1984: 250) argues that whereas there was a 'relative lack of colour consciousness among Europeans in earlier ages of expansion', in the nineteenth century, Europe 'racialized' the world. There was a general acceptance in Europe that in order to acquire statehood and be permitted to enter the European international society, political entities had to measure up to the European standard of civilization, despite the fact that, as Bull (1984c: 125) notes, the European states themselves could not live up to every aspect of the standard. However, it is also recognized in Bull and Watson that it is important not to overplay this line of argument because it has the effect of removing any sense of agency from non-European actors. As Howard notes, the Russians 'imitated' the Europeans because they wished to be able to compete more effectively and they then constituted a vanguard that others could follow. States like the Ottoman Empire, Japan and the Chinese Empire are shown to have followed the same route during the nineteenth century. Moreover, they also very quickly began to translate European and American international law textbooks and this helped them to assert their rights against the Europeans (Gong 1984b: 180, 181; Suganami 1984: 195; Bull 1984c: 121; Zhang 1991; Suzuki 2009).

Although it remains the case that the Europeans, but now more especially the Americans, are endeavouring to employ their own cultural norms and institutions to define the essential features of the global international society, these endeavours have always been challenged and there is no doubt that the contemporary structure of the global international institution is essentially multicultural in orientation. In Bull and Watson, there are two chapters that explore this phenomenon and they come to diametrically opposed conclusions. Bozeman (1984: 406) argues that the brief historical period when Western norms and institutions prevailed has gone, and this poses a fundamental challenge for European and American diplomacy, with Western diplomats having to function as they did before the nineteenth century 'in a world that has no common culture and no overarching political order, and that is no longer prepared to abide by western standards of international conduct'. By contrast, Dore (1984) argues that there is a need to distinguish between universal and idiosyncratic values and interests and he is convinced that despite the obvious existence of diversity there are universal values that underpin an emerging world culture. I shall return briefly to this dispute in the conclusion after comparing how the chapters in this book affect the argument presented in Bull and Watson.

A comparison

In the first place, Göl wishes to contest the paradox in Bull and Watson, presented by Naff (1984: 143), who claims that though a significant portion of the Ottoman Empire was '*in* Europe, it cannot be said to have been *of* Europe'. The central argument developed by Naff is that soon after the formation of the Ottoman Empire it began to pose a very serious threat to Christendom and one that persisted for several centuries, and during this initial phase the relationship with the Europeans was conducted very much on terms set by the Ottoman Empire. Like all world empires, the Ottomans believed that they had a duty to defend their civilization and this involved the protection of the worldwide Muslim community – the Dar al Islam. By the same token, they depicted the Europeans as inferiors and infidels, and their relationship with them was based on a state of war – the Dar al Harb. During this early phase, Naff argues that the empire effectively operated as an offshore balancer – using the expression coined later by Layne (1997). This role required the empire to pursue a strategy designed to promote disunity amongst the Europeans, which proved particularly effective during the Reformation in the sixteenth century. However, by the end of the century, the Habsburg Empire had found an effective counter, establishing an alliance with the shah of Iran and confronting the Ottomans with the possibility of war on two fronts. Over the next two centuries, the balance of power steadily shifted away from the Ottoman Empire and towards the Europeans. By the nineteenth century, the Ottomans had no alternative but to seek alliances with European states to protect them from the growing power of Russia.

The first crucial argument made by Göl is that Naff fails to acknowledge that for several centuries the Ottoman Empire lay at the heart of a regional system of global significance. It follows that Europe not only operated very much on the periphery of this system, but also that during this period the Ottoman Empire absorbed a substantial chunk of European territory. This image then gives rise to a second important argument, which suggests that the Ottoman Empire lay at the crossroads between the east and west of Eurasia and that the empire was open to influence from civilizations to the East and West, as well as from Africa, with the result that it displayed a distinct 'cultural hybridity'. This line of argument is seen to present the English School with an insuperable problem because of its assertion that the stability of the European international order was dependent upon a common European culture, and Bull, in particular, was doubtful if the cultural pluralism that characterizes contemporary international society can sustain a stable world order. This is now an issue that is being addressed by the English School (Buzan 2010) and, to be fair, Naff also acknowledges a 'process of cultural synthesis' between Europe and the Ottoman Empire. By the same token, Bull and Watson point specifically to the existence of an 'Arab Islamic system', as well as a more expansive nascent Eurasian international society, but neither of these insights is explored in any depth.

The failure in Bull and Watson to explore the Indian sub-continent is particularly surprising because key members of the English School were well aware of its importance. Watson (2009), for example, devoted a chapter to the Mauryan Empire illustrating a very early phase in the long process of evolution of the international society. Moreover, he also wrote a fascinating account of the complex political relations in India between Hindu and Muslim states in the fourteenth and fifteenth centuries (Watson 1964). Despite this interest, when it comes to the modern era, the Indian sub-continent, like the rest of the European imperial territory, simply drops out of the picture. Vigneswaran, by contrast, fully appreciates the importance of taking account of the Indian sub-continent in the broader process whereby the pluralistic order of distinct international societies began to converge. In his chapter in this volume he makes a seminal contribution to this literature.

By delving into the archives of the East India Company, Vigneswaran is able to throw light on the extraordinary way in which the English were able to insinuate themselves into the extant Indian international society. It was on the basis of these tentative foundations that the British were eventually able to establish themselves as the hegemonic actor in the Indian sub-continent during the course of the nineteenth century. However, his focus is on the opening phase of this development and his analysis illustrates very clearly the argument developed in the important study of imperial sovereignty by Benton (2010: 5), where she suggests that configuring imperial sovereignty involves 'pragmatic and theoretical questions arising from the entanglements of local legal politics and the challenges of interimperial contests'. When British traders first began to make serious efforts to trade with India in the late sixteenth century, maritime trade between Europe and India was already well entrenched. The trade had been initiated by the Portuguese but was increasingly being controlled at the European end by the Dutch and the French. At the Indian end, local rulers initially had a much better understanding of how questions of power and privilege operated within the established political systems that existed on the subcontinent, for example, the Mughal Empire in the north based on Agra and the kingdom of Golkonda in the south-east. These local rulers also proved adept at playing the Europeans off against each other as they sought trading concessions from these rulers through the courts. What Vigneswaran demonstrates, however, is that over time, local officials of the East India Company learned to exploit rivalries amongst imperial rivals, indigenous rulers as well as the local rulers. It was these local officials, often operating independently, who began to lay the basis for sovereign rights that were eventually taken over by the British state.

Although both the Ottoman Empire and the Indian sub-continent eventually converged with the European international society, albeit via rather different routes, as Neumann shows in this volume, the convergence with Russia occurred earlier and followed a different route again. Watson (1984b), who worked for some time in the British Embassy in Moscow, was well aware of many of the distinctive features of Russia's relations with Europe. He acknowledges, first, the importance of the successful invasion by the nomadic Tartars in the thirteenth century, although he also stresses that Christianity helped to preserve the identity of the Russians during the period that they were under the suzerain control of the Tartars. He also acknowledges the hostility of the Russian Orthodox Church to Catholicism with the result that the Russians were never considered part of Latin Christendom, even after the Roman Catholic Church and the Eastern Orthodox Church agreed to reunify in 1439. Instead, in 1448, the Russian Church established its independence from Constantinople. Once the Russians recovered their independence and began to pursue a policy of expansion in the sixteenth century, their distinctive identity emerged as they came into contact with their Catholic neighbours to the west, in Poland and Lithuania, as well as the Islamic Ottoman Empire to the south. Indeed, Watson depicts Russia as a power reaching out to both East and West. He focuses on the links that Russia forged with the security system in northern Europe and the importance of the Treaty of Nerchinsk (1689) with Manchu China. However, as the northern and southern security systems in Europe coalesced in the seventeenth century, so the Russians evinced substantial interest in exercising influence in the emerging European international society, and they emulated many of the developments that were taking place within this society. Nevertheless, Watson also points to a persistent tension in Russia between traditionalists and the Westernizers throughout this era.

A major problem with Watson's analysis, from Neumann's perspective, is that it fails to acknowledge the real significance of the connection between the Russians and the adjacent nomadic polities. For Western Europeans, the nomads were long perceived as alien and hostile, whereas for the Russians, they were neighbours with whom they developed an essentially reciprocal relationship that extended back to the eighth century. Neumann goes further and argues that when the Russians re-established their independence from the Tartars, a process of hybridization had taken place.[5] Nevertheless, the 250 years that the Russians operated under Tartar suzerainty are seen to have created a very distinctive hybrid polity that helps to explain some of the complexity of the process when Russia began to converge with the other European states. Elsewhere Neumann and Wigen (n.d., forthcoming) argue that Russia shares this characteristic with other states, such as Turkey, and much more attention needs to be paid to this phenomenon when endeavouring to understand the emergence of the contemporary international society and the severe reservations that the Western Europeans had about both countries.

From Neumann's perspective, the historical record indicates that Russia has much more in common with Japan and China than is generally supposed. All three polities were longstanding members of international societies that were very different from the international society that evolved in Europe. This common factor has to be taken into account when endeavouring to understand how the contemporary global international society came into being. Suganami (1984) certainly acknowledges that Japan was an important member of a distinctive international society, but also that it had relations with the European states that made maritime contact with the East in the sixteenth century. However, he argues that from 1600, in the aftermath of the warring states era, the Togukawa established a more centralized polity, although 'feudal' lords still effectively governed much of Japan. Indeed, the Ryūkyū Islands, in the south-west of Japan, continued to pay tribute to China and this situation was formalized in 1655 because the Tokugawa saw this link as a way of keeping open a trade conduit with China. However, Suganami argues that the fear that feudal lords might use external contacts to elevate their position led the Togukawa to favour an overall policy of 'seclusion' and for the next 200 years contact with European states, apart from Holland, was essentially prohibited.

Although Suganami acknowledges that there was an early initial period when the Japanese had some contact with the Europeans, followed by an era when they held the Europeans, even the Dutch, at arm's length, he devotes most of his attention to the mid-nineteenth century and after. Suzuki sees this as a mistake, because it draws our attention away from the earlier periods, when relations between Japan and Europe were largely established on Japanese terms. The effect, although not the intention, is not only to foster a Eurocentric interpretation of this specific relationship but also to draw attention away from the fact that Japan formed an important member of an international society, the origins of which can be traced back to an era before there was a European presence in the region. To counter this effect, Suzuki pays close attention to how the Dutch were required to relate to Japan during the so-called 'seclusion' era, which he also depicts as an era of transition away from the prevailing Sinocentric hierarchical order, with China at the apex of the hierarchy and the other polities in the international order occupying subordinate positions. Diplomatic relations among the polities were governed by elaborate rituals that were designed to maintain and strengthen the overall hierarchy. The nature of the Asian international order was somewhat different, therefore, to the order that was evolving in Europe, where the sense of hierarchy was slowly but surely giving way to an international order that embraced a concept of sovereign equality (Larkins 2010). The East Asian order, however, was also changing at this time, according to Suzuki, as the Japanese began to challenge its Sinocentric character, and to view itself as occupying the centre ground. The Europeans were then depicted as 'outer barbarians', but they were nevertheless co-opted in order to help to shore up and legitimize this Japanese view of the international order. To establish and maintain a trading connection, therefore, the Dutch had no alternative but to follow Japanese diplomatic protocol, which was designed to reinforce the existing hierarchical order, with the Dutch following procedures that identified the Japanese as occupying the pre-eminent position in the hierarchy, and the Dutch as 'outer barbarians'. This assessment, argues Suzuki, indicates the potential for international order in a multi-cultural world.

A similar conclusion can also be drawn from the Chinese case study. Gong (1984b) focuses initially on the fact that the encounter between the Chinese and Europeans from the sixteenth century onwards brought two very different standards of civilization into potential conflict. One area of dispute became very apparent in the seventeenth century when a disagreement between Jesuit and Dominican missionaries opened up about whether the Chinese who converted to Christianity could continue with their long-established Chinese rituals. The Jesuits argued that these rituals were social rites and were not in conflict with Christianity, and they themselves, impressed by aspects of Confucianism, adapted to a Confucian lifestyle – for example, wearing Chinese gowns. By the same token, the emperor accepted advice from the Jesuits on diplomacy, astronomy and military technology, with the Jesuits even taking responsibility for the emperor's observatory. In 1656, Pope Alexander VII decreed that Chinese practices were not incompatible with Christianity and in 1692 Emperor Kangxi issued an edict of toleration of Christianity. However, the Dominicans, and other Catholic orders, were appalled by these developments and from an early stage resisted them. Their view eventually prevailed and in 1705 Pope Clement XI prohibited Chinese Christians from practising Chinese rites, and in 1721 Emperor Kangxi banned Christian missions in China. Gong also points out that the divergent European responses to Chinese civilization extended to diplomacy and he notes how in 1689 in the Treaty of Nerchinsk between Russia and China, they agreed to reciprocate their established diplomatic rituals in their respective countries.

Zhang, in his chapter, explores this issue is much greater depth than Gong and aims to show, first, that there were sustained and substantial attempts by the Europeans to penetrate the Sinocentric world from the early sixteenth century onwards and, second, that the ability of the Europeans to make successful contact with the Chinese was highly dependent on their willingness to adapt to what Gong calls the Chinese standard of civilization. By endeavouring to find a Confucian–Christian synthesis, the Jesuits proved remarkably successful at promoting a Catholic presence in China, but as soon as the pope endeavoured to impose a European standard on the Chinese, the links were immediately severed. Zhang goes onto explore how both the Portuguese and Dutch offered to use their naval and military strength to support the emperor in an effort to gain a privileged trade position – although without much success. Nevertheless, the Europeans did succeed in insinuating themselves into the very important South-East Asian trading system, which extended far beyond the formal tribute system, but only on terms that were set by the Chinese and the Japanese. Finally, Zhang focuses on the European attempts to establish diplomatic relations with the Chinese. Again, there was no attempt to import European diplomatic protocol; instead, the Europeans displayed a willingness to work within the Sinocentric ceremonial traditions. Even so, the Europeans are seen to have had remarkably little success in finding ways to use diplomatic channels to open a conduit that would allow them to influence the emperor's views.

The final two case studies move us into very different and largely uncharted waters. Jones focuses on the Western hemisphere and surveys developments in North and South America. Although there are two chapters on these regions in Bull and Watson (1984a), neither touches on the period 1663 to 1820, which Jones refers to as the 'long eighteenth century'. He explores the complex interaction between three distinct groups: the indigenous population who originally populated the Americas; the Europeans who arrived at the end of the fifteenth century; and finally the creoles who were born in the Americas but whose forebears, or at least some of them, had been born in Europe. In essence, Jones provides an account of the ultimate military triumph of the emergent creole states over their European and indigenous competitors. This story has essentially been ignored in the International Relations literature, although clearly it must play a crucial role in any attempt to account for the emergence of the contemporary international society. It is clearly very different from any of the previous case studies: more tragic and more barbaric. Tragic because vast numbers of the indigenous population died as the result of diseases brought in from Europe – a process that persisted through to the twentieth century; and barbaric because the indigenous populations had to suffer slavery and genocide at the hands of the creoles and Europeans.

Jones can do no more than highlight some of the key features on this enormous canvas. He starts by noting that it took a long time before the Europeans and creoles were able to exercise very much control over either North or South America, and it is only in the latter third of the sixteenth century that it becomes possible to identify an emerging international society embracing indigenous and creole polities in Spanish South America, and a century later before this claim can be extended to North America and

Portuguese South America. Moreover, it took close to 400 years before the stubborn resistance by indigenous polities was finally overcome and the outlines of the contemporary states on the two continents began to take shape. This resistance was undoubtedly facilitated by a good deal of cross-cultural exchange, although Jones also notes that there was also a good deal of mutual incomprehension. However, the resistance was also assisted by a process of ethnogenesis taking place, with the North American Indians, for example, coming to see themselves as one people. Added to this there were also the rivalries between the European powers, on the one hand, and the divisions that opened up between the creoles who wanted independence and those who wished to maintain links with Europe, on the other. Finally, it is also the case that the Europeans endeavoured to mediate between the creoles and the indigenous people. As Hinderaker (1997) argues, this willingness to mediate partially explains why the Europeans found that their attempts to establish effective empires in the Americas proved to be so 'elusive'. By contrast, once the USA came into being, the federal government was able to establish a 'successful' empire by sponsoring the creoles, tacitly denying Indians' title to land, legitimizing the claims of settlers, and thereby creating a mechanism to give dependent territory the status of statehood.

Prior to the nineteenth century, European links with Africa were even more limited than those that developed in the Americas and they have been almost completely ignored in the International Relations literature. However, there is a chapter in Bull and Watson devoted to the topic, where Bull (1984c) draws a sharp contrast between North Africa, where relations with the Islamic world were quite strong, and Africa south of the Sahara, where, prior to the middle of the nineteenth century, with a few significant exceptions, the European contacts were restricted to the coast. On the basis of these coastal links, the Europeans followed in the footsteps of the Islamic world and established an even more substantial slave trade. Bull goes on to argue that there was a very complex and divergent set of polities across Africa, although they did not constitute a systemic or societal whole. While contacts tended to be very local, he does acknowledge a rich array of complex customs and institutions that allowed the African polities to interact with each other as well as with the Europeans on a sophisticated basis.

Quirk and Richardson provide a very much more detailed and more nuanced account of the relations that the Europeans had with Africa from the fifteenth century onwards, and they show how the Atlantic slave trade, in particular, depended upon the development of this multifaceted relationship and how it, in turn, had very significant consequences for the relationships that already existed amongst polities in Atlantic Africa. As in Asia, the Europeans had to find a way of inserting themselves into long-established institutions and patterns of behaviour in Africa and they often had no alternative but to accept that the Africans played the key role in determining the terms of their relationship. Slavery was a significant institution in Africa before the arrival of the Europeans, because labour was often a more scarce commodity than land. As a consequence, a victorious party in war would take people rather than land from the vanquished. Once the Europeans established a new demand for slaves, however, established institutions were transformed to satisfy this demand. It follows that although the Europeans only had a presence on the coast, with a few exceptions, and although they had to negotiate for slaves on terms set by the Africans, over time they had a transformative and devastating effect in many quarters of Africa. As Quirk and Richardson note, it is, therefore, a cruel irony that in the nineteenth century the Europeans went on to use the existence of slavery in Africa as a justification for colonization.

Conclusion

This chapter has had two main aims. The first was to suggest that the account of the expansion of the international society in Bull and Watson (1984a) is much more interesting and radical than is generally acknowledged, even by Bull and Watson themselves. Although at first sight it might seem that the book is concerned with the way that states around the globe entered a European-based international society during the course of the nineteenth and twentieth centuries and, indeed, Bull and Watson appear to associate themselves with what they depict as the 'standard' Eurocentric account, a closer investigation makes it clear that for Bull and Watson, prior to the nineteenth century the European international society operated in the context of a global environment made up of diverse regional international societies, each reflecting its own distinct cultural base. The second aim was to compare what the authors in this book have to say about these regionally based international societies with what the specific authors in Bull and Watson had to say about these societies. Broadly speaking, the overall message is the same in both sets of chapters. From the sixteenth century, at the latest, Europe was interacting with all of the international societies around the globe, but in every case, certainly in the initial centuries, very much on terms set by the non-European international societies. What we get in this book, however, is a much more extensive and nuanced account of this process. The Indian sub-continent and the Americas are included in the overall story; Russia and the Ottoman Empire, Japan and China are all depicted as members of non-European international societies that have all had a very major impact on the course of world history; and finally Europe's coastal links with Africa are seen to have spawned an Atlantic system that had enormous consequences across all three continents that the system brought together.

What both books reveal very clearly is that across four centuries the overall structure of the global international arena remained broadly unchanged, with a range of discrete regional international societies that were, albeit loosely, linked together by the contact made between them by the Europeans. The transformative change in the global arena occurred only during the course of the nineteenth century and it was precipitated by the self-conscious move by the European states to elevate their regional international society above the others – although this assessment needs to be further complicated by the emergence of the USA, which also played an important role in this development (see also Buzan and Lawson 2012a). As discussed in the first section of the chapter, in Bull and Watson this move is seen to involve a reversal of the way that the Europeans had been relating to these other regions. Instead of complying with the rules and institutions of these regions, the Europeans began to develop and impose their own rules and institutions. What this book does is to help clarify what international relations across the global arena looked like before the reversal began to take place. It demonstrates that when the Europeans made contact with other international societies, they were often willing – literally, in the case of China – to kowtow in order to establish contact with these non-European societies. However, by the same token, when these international societies wanted to make contact with the European international society, they had to make concessions and work within European norms and institutions. As becomes apparent in this book, prior to the nineteenth century, this was very much a two-way process, with countries like Russia and the Ottoman Empire accommodating to evolving European norms and institutions, but in other parts of the world it was the Europeans who had to make the accommodations. It is clear, therefore, that the convergence of the regional international societies was a complex process and it deserves much more study than it has received so far. Nevertheless, it is clear that whether or not it is possible to identify global values that are shared by all, the historical record does not support the unequivocal argument that cultural differences necessarily stand in the way of the emergence of a stable global political order.

Notes

1 This situation is now being rectified. See, for example, Hobson (2012).
2 The original project, to explore the evolution of international societies across world history was subsequently taken on and completed by Adam Watson. In an introduction to a reissue of the book, Buzan and Little (2009: ix) argue that the significance of this later book has not been fully acknowledged and in many ways it 'epitomises' the approach advocated by the British Committee.
3 Martin Wight died in 1972 but he had a major influence on both Bull and Watson.
4 For an attempt to examine the balance of power from a world historical perspective, see Kaufman, Little and Wohlforth (2007).
5 However, the links between sedentary and nomadic people is a Europe-wide phenomenon, so it is also necessary to acknowledge a degree of hybridization across Europe.

Bibliography

Abu-Lughod, J.L. (1989) *Before European Hegemony: The World System A.D. 1250–1350*, Oxford: Oxford University Press.

Acemoglu, D., Johnson, S. and Robinson, J.A. (2002) 'Reversal of Fortune: Geography and Institutions in the Making of the Modern World Income Distribution', *Quarterly Journal of Economics* 117(4): 1231–94.

——(2005) 'The Rise of Europe: Atlantic Trade, Institutional Change, and Economic Growth', *American Economic Review* 95(3): 546–79.

Adanır, F. (2005) 'Turkey's Entry into the Concert of Europe', *European Review* 13(3): 395–417.

Adas, M. (1989) *Machines as the Measure of Men: Science Technology, and Ideologies of Western Dominance*, Ithaca, NY: Cornell University Press.

Agor, W.H. (1972) 'Latin American Inter-State Politics: Patterns of Cooperation and Conflict', *Inter-American Economic Affairs* 26(2): 19–33.

Alabi, Adetayo (2008) 'Introduction: Africa in a Global Age', *The Global South* 2(2): 2–3.

Albrecht-Carrie, R. (1958) *A Diplomatic History of Europe since the Congress of Vienna*, London: Methuen & Co. Ltd Press.

Allsen, T.T. (1987) *Mongol Imperialism: The Politics of the Great Qan Möngke in China, Russia, and the Islamic Lands, 1251–1259*, Berkeley: University of California Press.

——(1997) *Commodity and Exchange in the Mongol Empire: A Cultural History of Islamic Textiles*, Cambridge: Cambridge University Press.

Alpern, S. (1995) 'What Africans Got for their Slaves: A Master List of European Trade Goods', *History in Africa* 22: 5–43.

Amitai-Preiss, R. and Morgan, D. (eds) (1999) *The Mongol Empire and its Legacy*, Leiden: Brill.

Anderson, F. (2000) *Crucible of War: The Seven Years' War and the Fate of Empire in British North America, 1754–1766*, London: Faber.

Anghie, A. (2004) *Imperialism, Sovereignty and the Making of International Law*, Cambridge: Cambridge University Press.

——(2006) 'The Evolution of International Law: Colonial and Postcolonial Realities', *Third World Quarterly* 27(5): 739–53.

Anonymous (2011) 'The Sun Shines Bright: The Continent's Impressive Growth Looks Likely to Continue', *The Economist*, 3 December, www.economist.com/node/21541008/ (accessed 2 February 2013).

Arano, Y. (1988) *Kinsei nippon to higashi ajia*, Tokyo: Tōkyō daigaku shuppankai.

Armaoğlu, F. (1987) *20. Yüzyıl Siyasi Tarihi: 1914–1980*, Ankara: Türkiye İş Bankası Yayınları.

Armitage, D. (2000) *The Ideological Origins of the British Empire*, Cambridge: Cambridge University Press.

Arrighi, G., Hamashita, T. and Seldon, M. (eds) (2003) *The Resurgence of East Asia: 500, 150 and 50 Years Perspectives*, London: Routledge.

Atwell, W. (1998) 'Ming China and the Emerging World Economy, c. 1470–1650', in Denis Twitchett and Frederick W. Mote (eds) *Cambridge History of China, Vol. 8, The Ming Dynasty, 1368–1644, Part 2*, Cambridge: Cambridge University Press, 376–416.

Austen, G. (2008) 'The "Reversal of Fortune" Thesis and the Compression of History: Perspectives from African and Comparative Economic History', *Journal of International Development* 20: 996–1027.

Austen, R. (2010) *Trans-Saharan Africa in World History*, Oxford: Oxford University Press.

Austen, R.A. and Derrick, J. (1999) *Middlemen of the Cameroons Rivers: The Duala and their Hinterland, c.1600–c.1900*, Cambridge: Cambridge University Press.

Axtell, J. (1978) 'The Ethnohistory of Early America: A Review Essay', *The William and Mary Quarterly* 35(1): 110–44.

——(1988) *After Columbus: Essays in the Ethnohistory of Colonial North America*, New York: Oxford University Press.

Backus, O.P. (1957) *Motives of the West Russian Nobles in Deserting Lithuania for Moscow, 1377–1514*, Lawrence, KS: University of Kansas Press.

Bailyn, B. (1986) *The Peopling of British North America: An Introduction*, New York: Alfred A. Knopf.

——(2005) *Atlantic History: Concepts and Contours*, Cambridge, MA: Harvard University Press.

Barfield, Thomas J. (1989) *The Perilous Frontier: Nomadic Empires and China*, ed. Charles Tilly, Cambridge, MA: Blackwell.

Barkawi, T. and Laffey, M. (2002) 'Retrieving the Imperial: Empire and International Relations', *Millennium* 31(1): 109–27.

Barraclough, G. (ed.) (1999) *The Times History of the World*, new edn, ed. by Richard J. Overy, London: Times Books.

Bartlett, R. (1994) *The Making of Europe: Conquest, Colonization and Cultural Change 950–1350*, London: Penguin Books.

Basil, N. (2011) 'The Roots of Afropessimism: The British Invention of the "Dark Continent"', *Critical Arts* 25(3): 377–96.

Bassett, D.K. (1998) 'Early English Trade and Settlement in Asia, 1602–90', in P.J.N. Tuck (ed.) *The East India Company, 1600–1858*, New York: Routledge, 1998.

Bayart, J. (2009) *The State in Africa: The Politics of the Belly*, Cambridge: Polity.

Bayly, Christopher (2004) *The Birth of the Modern World: Global Connections and Comparisons, 1780–1914*, Oxford: Blackwell.

Benjamin, T. (2009) *The Atlantic World: Europeans, Africans, Indians and their Shared History, 1400–1900*, Cambridge: Cambridge University Press.

Bennison, Amira K. (2009) 'The Ottoman Empire and its Precedents from the Perspective of English School Theory', in B. Buzan and A. Gonzalez-Pelaez (eds) *International Society and the Middle East: English School Theory at the Regional Level*, London: Palgrave Macmillan.

Benton, L. (2010) *A Search for Sovereignty: Law and Geography in European Empires, 1400–1900*, Cambridge: Cambridge University Press.

Berger, M.T. (1990) 'From Commerce to Conquest: The Dynamics of British Mercantile Imperialism in Eighteenth-Century Bengal, and the Foundation of the British Indian Empire', *Bulletin of Concerned Asian Scholars* 22(1): 44–62.

Bethell, L. (ed.) (1984) *The Cambridge History of Latin America*, Cambridge: Cambridge University Press.

Bielenstein, H. (2005) *Diplomacy and Trade in the Chinese World 589–1276*, Leiden and Boston, MA: Brill.

Birmingham, D. (1966) *Trade and Conflict in Angola: The Mbundu and their Neighbours Under the Influence of the Portuguese 1483–1790*, Oxford: Clarendon Press.

Bitterli, Urs (1989) *Cultures in Conflict: Encounters between European and Non-European Cultures, 1492–1800*, trans. Richie Robertson, Cambridge: Polity Press.

Blackburn, R. (2012) *The American Crucible: Slavery, Emancipation and Human Rights*, London: Verson.

Boccara, G. (1999) 'Etnogénesis mapuche: resistencia y restucturación entre los indígenas del centro-sur de Chile (siglos XVI–XVIII)', *Hispanic American Historical Review* 79(3): 425–61.

Bolin, W.H. (1992) 'The Transformation of South America's Borderlands', in L.A. Herzog (ed.) *Changing Boundaries in the Americas: New Perspectives on the U.S.–Mexican, Central American, and South American Borders*, San Diego: University of California at San Diego.

Boogaart, E. van den (1992) 'The Trade between Western Africa and the Atlantic World, 1600–690: Estimates of Trends in Composition and Value', *The Journal of African History* 33(3): 369–85.

Bowden, B. (2004) 'In the Name of Progress and Peace: The "Standard of Civilization" and the Universalizing Project', *Alternatives* 29(1): 43–68.

——(2009) *The Empire of Civilization: A Story about Making History and Influencing Peoples*, Chicago, IL: University of Chicago Press.

Bowen, H.V. (1988) 'A Question of Sovereignty? The Bengal Land Revenue Issue, 1765–67', *Journal of Imperial and Commonwealth History* 16(2): 155–76.

——(1989) 'Investment and Empire in the Later Eighteenth Century: East India Stockholding, 1756–91', *The Economic History Review* 42(2): 186–206.

——(1991) *Revenue and Reform: The Indian Problem in British Politics, 1757–1773*, Cambridge: Cambridge University Press.

Boxer, C.R. (1969) *The Portuguese Seaborne Empire, 1415–1825*, London: Hutchinson & Co. Ltd.

——(1984) 'When the Twain First Met: European Conceptions and Misconceptions of Japan, Sixteenth-Eighteenth Centuries', *Modern Asian Studies* 18(4): 531–40.

Bozeman, A. (1984) 'The International Order in a Multicultural World', in H. Bull and A. Watson (eds) *The Expansion of International Society*, Oxford: Clarendon Press.

Bradshaw, B. and Morrill, J.S. (eds) (1996) *The British Problem, c.1534–1707: State Formation in the Atlantic Archipelago*, Basingstoke: Macmillan.

Branch, J. (2012) '"Colonial Reflection" and Territoriality: The Peripheral Origins of Sovereign Statehood', *European Journal of International Relations* 18(2): 277–97.

Brand, D.D. (1941) 'A Brief History of Araucanian Studies', *New Mexico Anthropologist* 5(2): 19–35.

Brantlinger, P. (1985) 'Victorians and Africans: The Genealogy of the Dark Continent', *Critical Inquiry* 12(1): 166–203.

Braudel, F. (1984) *Civilisation and Capitalism: 15th–18th Century, Vol. III: The Perspective of the World*, London: Collins.

——(1994) *A History of Civilizations*, trans. Richard Mayne, London: Allen Lane.

British Library Archives (1626a) *Official Correspondence (OC), Factory Records, Masulipatam, 'Letter from Armagon to Batavia: Signed Thomas Johnson and Thomas Mille'*, 26 January, 48.

——(1626b) *OC, Factory Records, Masulipatam, 'Letter from Armagon to Batavia: Signed Thomas Johnson and Beaverly'*, 19 April.

——(1626c) *OC – Factory Records, Masulipatam, 'Letter from Masulipatam to Batavia: Signed Thomas Mille'*, 3 June [transcriber's notes say this letter was actually written by Thomas Johnson].

——(1628) *OC, Factory Records, Masulipatam, 'Letter to the King of Golconda: signed George Brewen, Thomas Johnson and Lawrence Henley'* [attached to a consultation of 16 September 1628].

——(1640) *OC, Factory Records Miscellaneous 9, 'Letter to East India Company in England: Signed W. Fremlen, Fra. Breton, Ben Robinson, Jn. Wylde'*, 29 December.

——(1642) *OC, Factory Records Miscellaneous 9, Document E 'Letter to Bantam: Signed Thomas Cogan, Henry Greenhill and John Browne'*, 20 September.

——(1645) *OC, Factory Records Miscellaneous 9, Document K. 'Report of a Committee appointed to hear Cogan's case: Wm Methwold Deputy; Ashwell; Mr Massingberd Treasurer; Mr Holloway; Mr Wilson'*, 13 May.

——(1661) *The East-India Companies Charter, Granted by the Kings Most Excellent Majesty Charles the Second, under the Great Seal of England*, 3 April.

——(n.d.) *OC, Factory Records Miscellaneous 9, Document A 'Firman granted by Damela Vintutedro Naiqe to Francis Day' – Chief of Armagon factory on behalf of the East India Company'.*

——(n.d.) *OC, Factory Records Miscellaneous 9, Document F 'Firman to Thomas Ivy'.*

——(n.d.) *OC, Factory Records Miscellaneous 9, 'Anonymous prefatory Remarks'.*

——(n.d.) *OC, Factory Records Miscellaneous 9, Document C, 'Letter from Masulipatam to Company in England'.*

Brooks, G. (2003) *Eurafricans in Western Africa: Commerce, Social Status, Gender and Religious Observance from the Sixteenth to the Eighteenth Century*, Oxford: James Curry.

Brown, C. (1995) 'International Theory and International Society: The Viability of the Middle Way?' *Review of International Studies* 21(2): 183–96.

Brownlie, Ian (1984) 'The Expansion of International Society: The Consequences for International Law', in H. Bull and A. Watson (eds) *The Expansion of International Society*, Oxford: Clarendon Press.

Bryant, G. (1998) 'Officers of the East India Company's Army in the Days of Clive and Hastings', in Patrick J.N. Tuck (ed.) *The East India Company, 1600–1858*, New York: Routledge.

Bukh, A. (2010) *Japan's National Identity and Foreign Policy: Russia as Japan's 'Other'*, London: Routledge.

Bull, H. (1984a) 'Revolt Against the West', in H. Bull and A. Watson (eds) *The Expansion of International Society*, Oxford: Clarendon Press.

——(1984b) 'European States and African Political Communities', in H. Bull and A. Watson (eds) *The Expansion of International Society*, Oxford: Clarendon Press.

——(1984c) 'The Emergence of a Universal International Society', in H. Bull and A. Watson (eds) *The Expansion of International Society*, Oxford: Clarendon Press.

——(1995 [1977]) *The Anarchical Society: A Study of Order in World Politics*, Basingstoke: Macmillan.

Bull, H. and Watson, A. (1984a) *The Expansion of International Society*, Oxford: Clarendon Press.

——(1984b) 'Introduction', in H. Bull and A. Watson (eds) *The Expansion of International Society*, Oxford: Clarendon Press.

——(1984c) 'Conclusions', in H. Bull and A. Watson (eds) *The Expansion of International Society*, Oxford: Clarendon Press.

Butterfield, H. (1931) *The Whig Interpretation of History*, London: G. Bell & Sons.

Buzan, B. (1993) 'From International System to International Society: Structural Realism and Regime Theory Meet the English School', *International Organization* 47(3): 327–52.

——(2004) *From International to World Society? English School Theory and the Social Structure of Globalisation*, Cambridge: Cambridge University Press.

——(2010) 'Culture and International Society', *International Affairs* 86(1): 1–25.

Buzan, B. and Lawson, G. (2012a) 'The Global Transformation: The Nineteenth Century and the Making of Modern International Relations', *International Studies Quarterly*, forthcoming.

——(2012b) 'Rethinking Benchmark Dates in International Relations', *European Journal of International Relations*, forthcoming.

Buzan, B. and Little, R. (2000) *International Systems in World History: Remaking the Study of International Relations*, Oxford: Oxford University Press.

——(2009) 'Introduction to the 2009 Reissue', in Adam Watson, *The Evolution of International Society: A Comparative Historical Analysis*, London: Routledge.

——(2010) 'The Evolution of International Society', in *ISA Compendium: The English School*, New York: Blackwell.

Cain, P.J. and Hopkins, A.G. (1980) 'The Political Economy of British Expansion Overseas, 1750–1914', *The Economic History Review* 33(4): 463–90.

——(1986) 'Gentlemanly Capitalism and British Expansion Overseas I: The Old Colonial System, 1688–1850', *The Economic History Review* 39(4): 501–25.

——(1987) 'Gentlemanly Capitalism and British Expansion Overseas II: New Imperialism, 1850–1945', *The Economic History Review* 40(1): 1–26.

——(2001) *British Imperialism, 1688–2000*, 2nd edn, New York: Longman.

Calderisi, R. (2007) *The Trouble with Africa: Why Foreign Aid Isn't Working*, New Haven, CT: Yale University Press.

Campbell, G., Miers, S. and Miller, J.C. (eds) (2007) *Women and Slavery: Africa, the Indian Ocean World, and the Medieval North Atlantic*, Athens: Ohio University Press.

——(eds) (2009) *Children in Slavery Through the Ages*, Athens: Ohio University Press.

Candido, M. (2006) *Enslaving Frontiers: Slavery, Trade and Identity in Benguela, 1780–1850*, York University, PhD dissertation.

Centeno, M.A. (1997) 'Blood and Debt: War and Taxation in Nineteenth-Century Latin America', *American Journal of Sociology* 102(6): 1565–605.

Chabal, Patrick and Daloz, Jean-Pascal (1999) *Africa Works: Disorder as a Political Instrument*, Oxford: James Currey.

Chaudhuri, K.N. (1985) *Trade and Civilization in the Indian Ocean: An Economic History from the Rise of Islam to the 1750*, Cambridge: Cambridge University Press.

——(1986) 'The English East India Company and its Decision-Making', in John Harrison (ed.) *East India Company Studies: Papers Presented to Professor Sir Cyril Philips*, Hong Kong: Asian Research Service.

Cherniavsky, M. (1970) 'Khan or Basileus: An Aspect of Russian Medieval Political Theory', in M. Cherniavsky (ed.) *The Structure of Russian History*, New York: Random House, 195–211.

Chrysos, E. (1992) 'Byzantine Diplomacy, A.D. 300–800: Means and End', in J. Shepard and S. Franklin (eds) *Byzantine Diplomacy. Papers from the Twenty-fourth Spring Symposium of Byzantine Studies, Cambridge, March 1990*, Aldershot: Variorum.

Çiçek, K. (ed.) (2001) *Pax Ottomana: Studies in Memoriam Prof. Dr. Nejat Göyünç (1925–2001)*, Haarlem: Ankara.

Clapham, C. (1996) *Africa and the International System: The Politics of State Survival*, Cambridge: Cambridge University Press.

Clark, I. (2005) *Legitimacy in International Society*, Oxford: Oxford University Press.

Cohen, P.A. (1984) *Discovering History in China: American Historical Writing on the Recent Chinese Past*, New York: Columbia University Press.

Cohen, R. (2008) *Global Diasporas: An Introduction*, London: Routledge.

Cohn, B.S. (1987) *Anthropologist among the Historians and Other Essays*, Delhi: Oxford University Press.

Coles, P. (1968) *The Ottoman Impact on Europe*, New York: Harcourt, Brace and World Publication.

Collier, P. (2007) *The Bottom Billion: Why the Poorest Countries are Failing and What Can Be Done About It*, Oxford: Oxford University Press.

Comaroff, J. and Comaroff, J. (2007) 'Law and Disorder in the Post-colony', *Social Anthropology/Anthropologie Sociale* 15(2): 133–52.

Conceição, P. (2011) 'The African Moment: On the Brink of a Development Breakthrough', Working Paper, 2011-001, United Nations Development Programme, June.

Connah, G. (2001) *African Civilizations: An Archaeological Perspective*, Cambridge: Cambridge University Press.

Cooper, F. and Burbank, J. (2011) *Empires in World History: Power and the Politics of Difference*, Princeton, NJ: Princeton University Press.

Cornelissen, S., Cheru, F. and Shaw, T. (2011) *Africa and International Relations in the 21st Century*, Houndmills: Palgrave.

Cox, R.W. (with Sinclair, T.J.) (1996) *Approaches to World Order*, Cambridge: Cambridge University Press.

Crawford, Neta (2002) *Argument and Change in World Politics*, Cambridge: Cambridge University Press.

Curtin, P. (1964) *The Image of Africa: British Ideas and Action, 1780–1850*, Madison: University of Wisconsin Press.

——(1975) *Economic Change in Precolonial Africa*, Madison: University of Wisconsin Press.

Curto, J. (2004) *Enslaving Spirits: The Portuguese-Brazilian Alcohol Trade at Luanda and its Hinterland, c. 1550–1830*, Leiden: Brill.

Curtoni, R., Lazzari, A. and Lazzari, M. (2003) 'Middle of Nowhere: A Place of War Memories, Commemoration, and Aboriginal Re-Emergence (La Pampa, Argentina)', *World Archaeology* 35(1): 61–78.

Daaku, K. (1970) *Trade and Politics on the Gold Coast 1600–1720*, Oxford: Oxford University Press.
D'Altroy, T. (2002) *The Incas*, Malden, MA and Oxford: Blackwell.
Darby, H.C. and Fullard, H. (1970) *The New Cambridge Modern History Atlas* (Volume 14 of *The New Cambridge Modern History*), Cambridge: Cambridge University Press.
Darwin, J. (2008) *After Tamerlane: The Global History of Empire Since 1405*, London: Bloomsbury.
da Silva, F.R. (2011) *Dutch and Portuguese in Western Africa: Empires, Merchants and the Atlantic System, 1580–1674*, Leiden: Brill.
Dawson, C. (ed.) (1955) *The Mongol Mission*, New York: Sheed & Ward.
de Carvalho, B., Leira, H. and Hobson, J.M. (2011) 'The Big Bangs of IR: The Myths that your Teachers Still Tell You about 1648 and 1919', *Millennium* 39(3): 735–58.
de Rachewiltz, I. (1971) *Papal Envoys to the Great Khans*, London: Faber & Faber.
de Rubruquis, G. (1990 [1255]) *The Mission of Friar William of Rubruck: His Journies to the Court of the Great Khan Möngke 1253–1255*, ed. P. Jackson and D. Morgan, London: Hakluyt.
Diène, D. (ed.) (2001) *From Chains to Bonds: The Slave Trade Revisited*, Paris: UNESCO.
Diouf, S.A. (2003) 'Introduction', in S.A. Diouf (ed.) *Fighting the Slave Trade: West African Strategies*, Athens: Ohio University Press.
Donelan, M. (1984) 'Spain and the Indies', in H. Bull and A. Watson (eds) *The Expansion of International Society*, Oxford: Clarendon Press.
Dong, Shaoxin and Huang, Yinong (2009) 'Chongzheng nianjian zhaomu pubing xinkao', *Lishi yanjiu* 5: 65–86.
Dore, R. (1984) 'Unity and Diversity in Contemporary World Culture', in H. Bull and A. Watson (eds) *The Expansion of International Society*, Oxford: Clarendon Press.
Dowd, G. (1991) *A Spirited Resistance: The North American Indian Struggle for Unity, 1745–1815*, Baltimore, MD: The Johns Hopkins University Press.
Dower, J.W. (1975) 'E.H. Norman, Japan and the Uses of History', in J.W. Dower (ed.) *Origins of the Modern Japanese State: Selected Writings of E.H. Norman*, New York: Pantheon Books.
Doyle, M. (1986) *Empires*, Ithaca, NY: Cornell University Press.
Du, Shiran and Han, Qi (1993) '17,18 shiji faguo yesu huishi dui zhongguo kexue de gongxian', *Kexue dui shehui de yingxiang* 167: 55–64.
Dunn, Kevin and Shaw, Timothy (eds) (2001) *Africa's Challenge to International Relations Theory*, Basingstoke: Palgrave.
Dunne, T. (1998) *Inventing International Society: A History of the English School*, Basingstoke: Palgrave Macmillan.
Eisenstadt, S.N. (2000) 'Multiple Modernities', *Daedalus* 129(1): 1–29.
——(ed.) (2002) *Multiple Modernities*, Piscataway, NJ: Transaction Publishers.
Elbl, I. (1991) 'The Horse in Fifteenth-Century Senegambia', *The International Journal of African Historical Studies* 24(1): 85–110.
Elias, N. (1978) *The History of Manners: The Civilizing Process*, Vol. 1, New York: Pantheon Books.
——(1982) *Power and Civility: The Civilizing Process*, Vol. 2, New York: Pantheon Books.
Ellis, H. (1824) *Original Letters illustrative of English History; including numerous royal letters: from Autographs in the British Museum and one or two other Collections*, Vol. III, London: Harding, Triphook, and Lepard.

Eltis, D. (2000) *The Rise of African Slavery in the Americas*, Cambridge: Cambridge University Press.

Eltis, D. and Jennings, L. (1988) 'Trade between Western Africa and the Atlantic World in the Precolonial Era', *American Historical Review* 93: 936–59.

Eltis, D., Lovejoy, P. and Richardson, D. (1999) 'Slave-Trading Ports: An Atlantic-Wide Perspective, 1676–1832', in R. Law and S. Stricktrodt (eds) *Ports of the Slave Trade (Bights of Benin and Biafra)*, Stirling: Centre of Commonwealth Studies.

Eltis, D. and Richardson, D. (eds) (2008) *Extending the Frontiers: Essays on the New Transatlantic Slave Trade Database*, New Haven, CT: Yale University Press.

——(2010) *Atlas of the Transatlantic Slave Trade*, New Haven, CT: Yale University Press.

Englebert, P. (2009) *Africa: Unity, Sovereignty and Sorrow*, London: Lynne Rienner.

Escobar, A. (1995) *Encountering Development: The Making and Unmaking of the Third World*, Princeton, NJ: Princeton University Press.

Evans, E.W. and Richardson, D. (1995) 'Hunting for Rents: The Economics of Slaving in Pre-Colonial Africa', *Economic History Review* 48(4): 678–99.

Fairbank, J.K. (ed.) (1968) *The Chinese World Order*, Cambridge, MA: Harvard University Press.

Falola, T. and Jennings, C. (eds) (2003) *Sources and Methods in African History: Spoken, Written, Unearthed*, Rochester, NY: University of Rochester Press.

Falola, T. and Roberts, K. (eds) (2008) *The Atlantic World, 1450–2000*, Bloomington: University of Indiana Press.

Faron, L.C. (1960) 'Effects of Conquest on the Araucanian Picunche during the Spanish Colonization of Chile, 1536–1635', *Ethnohistory* 7(3): 239–307.

Faroqhi, S. (1994) 'Part II: Crisis and Change, 1590–1699', in S. Faroqhi *et al.*, *An Economic and Social History of the Ottoman Empire*, Vol. 2: 1600–1914, Cambridge: Cambridge University Press.

——(1995) 'Iktisat Tarihi (1500–1600)', in S. Aksin (ed.) *Turkiye Tarihi, Vol. 2: Osmanli Devleti, 1300–1600*, Istanbul: Cem Yayinevi.

Feiling, K.G. (1954) *Warren Hastings*, London, New York: Macmillan, St Martin's Press.

Fennell, J.L.I. (1983) *The Crisis of Medieval Russia 1200–1304*, London: Longman.

Ferguson, N. (2011) 'The West and the Rest: The Changing Global Balance of Power in Historical Perspective', presentation at Chatham House, transcript at www.chathamhouse.org/sites/default/files/19251_090511ferguson.pdf (19 January 2013).

Fernández-Arnesto, F. (2003) *The Americas: The History of a Hemisphere*, London: Phoenix.

Ferreira, R. (2012) *Cross-Cultural Exchange in the Atlantic World: Angola and Brazil during the Era of the Slave Trade*, Cambridge: Cambridge University Press.

Fidler, D.P. (2000) 'A Kinder, Gentler System of Capitulations? International Law, Structural Adjustment Policies, and the Standard of Liberal, Globalized Civilization', *Texas International Law Journal* 35(3): 387–414.

——(2001) 'The Return of the Standard of Civilization', *Chicago Journal of International Law* 2: 137–57.

Finkel, C. (2005) *Osman's Dreams: The Story of the Ottoman Empire, 1300–1923*, London: John Murray Press.

Fisher, R. (1996) 'The Northwest from the Beginning of Trade with Europeans to the 1880s', in B.G. Trigger and W.E. Washburn (eds) *The Cambridge History of the Native Peoples of the Americas* (Vol.1: North America, Part 2), Cambridge: Cambridge University Press.

Foster, W. (1902) *The Founding of Fort St. George, Madras*, London: Eyre and Spottiswood.

Fowler, L. (1996) 'The Great Plains from the Arrival of the Horse to 1885', in B.G. Trigger and W.E. Washburn (eds) *The Cambridge History of the Native Peoples of the Americas* (Vol.1: North America, Part 2), Cambridge: Cambridge University Press.

Frank, A.G. (1998) *ReOrient: Global Economy in the Asian Age*, Berkeley: University of California Press.

Franklin, S. (2001) 'Pre-Mongol Rus: New Sources, New Perspectives?' *Russian Review* 60(4): 465–73.

Franklin, S. and Shepard, J. (1996) *The Emergence of Rus 750–1200*, London: Longman.

Friedman, G. and Lebard, M. (1991) *The Coming War with Japan*, New York: St Martin's Press.

Fukuyama, F. (1989) 'The End of History?' *The National Interest* 16: 3–18.

Galenson, D.W. (1986) *Traders, Planters, and Slaves: Market Behaviour in Early English America*, Cambridge: Cambridge University Press.

Geanakoplos, D. (1976) *Interaction of Sibling Byzantine and Western Cultures in the Middle Ages and Renaissance*, New Haven, CT: Yale University Press.

Gemery, H.A., Hogendorn, J. and Johnson, M. (1990) 'Evidence on English/African Terms of Trade in the Eighteenth Century', *Explorations in Economic History* 27(2): 157–77.

Gibson, C. (1964) *The Aztecs under Spanish Rule: A History of the Indians of the Valley of Mexico*, Stanford, CA: Stanford University Press.

Gillard, D. (1984) 'British and Russian Relations with Asian Governments in the Nineteenth Century', in H. Bull and A. Watson (eds) *The Expansion of International Society*, Oxford: Clarendon Press.

Gilpin, R. (1981) *War and Change in World Politics*, Cambridge: Cambridge University Press.

Gledhill, A. (1964) *The Republic of India: The Development of Its Laws and Constitution*, 2nd edn, London: Stevens & Sons.

Gleig, G.R. (1841) *Memoirs of the Life of the Right Hon. Warren Hastings, First Governor-General of Bengal: Compiled from Original Papers*, London: R. Bentley.

Gökçe, C. (1979) *Kafkasya ve Osmanlı İmparatorluğunun Kafkasya Siyaseti*, İstanbul: Has Kutulmuş Matbaası.

Goldfrank, D. (2000) 'Muscovy and the Mongols: What's What and What's Maybe', *Kritika: Explorations in Russian and Eurasian History, New Series* 1(2): 259–66.

Gong, G.W. (1984a) *The Standard of 'Civilization' in International Society*, Oxford: Clarendon Press.

——(1984b) 'China's Entry into International Society', in H. Bull and A. Watson (eds) *The Expansion of International Society*, Oxford: Clarendon Press.

Goodman, G.K. (2000) *Japan and the Dutch 1600–1853*, Richmond: Curzon Press.

Goody, J. (2010) *The Eurasian Miracle*, Cambridge: Polity Press.

Green, M.D. (1996) 'The Expansion of European Colonization to the Mississippi Valley, 1780–1880', in B.G. Trigger and W.E. Washburn (eds) *The Cambridge History of the Native Peoples of the Americas* (Vol.1: North America, Part 1), Cambridge: Cambridge University Press.

Gregson, R.E. (1969) 'The Influence of the Horse on Indian Cultures of Lowland South America', *Ethnohistory* 16(1): 33–50.

Grekov, B.D. and Yakubovskiy, A.Y. (1950) *Zolotaya Orda i ee padenie*, Moscow: Izdatel'stvo Akademii Nauk.

Grenier, J. (2005) *The First American Way of War: American War Making on the Frontier*, Cambridge: Cambridge University Press.

Grovogui, S. (1996) *Sovereigns, Quasi Sovereigns, and Africans*, Minneapolis: University of Minnesota Press.

Gu, Weimin (2002) 'Cooperation and Contradiction: Portugal and the Holy See in the Ecclesiastical Affairs of China in the 17th and 18th Centuries', *Revista de Cultura* 2: 90–95.

Halecki, O. (1952) 'Imperialism in Slavic and East European History', *American Slavic and East European History Review* 11(1): 1–26.

Hall, S. (1992) 'The West and the Rest: Discourse and Power', in S. Hall and B. Gieben (eds) *Formations of Modernity*, Cambridge: Polity.

Halliday, Fred (2009) 'The Middle East and Conceptions of "International Society"', in B. Buzan and A. Gonzalez-Pelaez (eds) *International Society and the Middle East: English School Theory at the Regional Level*, London: Palgrave Macmillan.

Halper, S. (2010) *The Beijing Consensus: How China's Authoritarian Model will Dominate the Twenty-First Century*, New York: Basic Books.

Halperin, C.J. (1983) 'Russia in the Mongol Empire in Comparative Perspective', *Harvard Journal of Asiatic Studies* 43(1): 239–61.

——(1987) *Russia and the Golden Horde: The Mongol Impact on Medieval Russian History*, Bloomington: Indiana University Press.

——(2000) 'Muscovite Political Institutions in the 14th Century', *Kritika: Explorations in Russian and Eurasian History, New Series* 1(2): 237–57.

Hamashita, T. (2008) *China, East Asia and the Global Economy: Regional and Historical Perspectives*, ed. Linda Grove and Mark Selden, London: Routledge.

Hamnett, B. (1999) *A Concise History of Mexico*, Cambridge: Cambridge University Press.

Hanioglu, M.S. (2008) *A Brief History of the Late Ottoman Empire*, Princeton, NJ: Princeton University Press.

Harling, P. (1996) *The Waning of 'Old Corruption': The Politics of Economical Reform in Britain, 1779–1846*, Oxford: Oxford University Press.

Harms, R. (1981) *River of Wealth, River of Sorrow: The Central Zaire Basin in the Era of the Slave and Ivory Trade, 1500–1891*, New Haven, CT: Yale University Press.

Harris, O. (1995) '"The Coming of the White People": Reflections on the Mythologisation of History in Latin America', *Bulletin of Latin American Research* 14(1): 9–24.

Hawthorn, W. (2010) *From Africa to Brazil: Culture, Identity, and an Atlantic Slave Trade, 1600–1830*, Cambridge: Cambridge University Press.

Hemming, J. (1978) *Red Gold: The Conquest of the Brazilian Indian*, London: Macmillan.

Herbst, J. (2000) *States and Power in Africa: Comparative Lessons in Authority and Control*, Princeton, NJ: Princeton University Press.

Hevia, James L. (1995) *Cherishing Men from Afar: Qing Guest Ritual and the Macartney Mission of 1793*, Durham, NC: Duke University Press.

Heywood, C. (2002) 'Review of *The Ottoman Empire, 1300–1650: The Structure of Power*', *Reviews in History* 431, www.history.ac.uk/reviews/review/431#response (accessed 6 May 2012).

Heywood, L. and Thornton, J. (2007) *Central Africans, Atlantic Creoles, and the Foundation of the Americas, 1585–1660*, Cambridge: Cambridge University Press.

Hinderaker, E. (1997) *Elusive Empires: Constructing Colonialism in the Ohio Valley, 1673–1800*, Cambridge: Cambridge University Press.

Hindess, B. (2000) ‘Citizenship in the International Management of Populations’, *American Behavioural Scientist* 43(9): 1486–97.

Hirono, M. (2008) *Civilizing Missions: International Religious Agencies in China*, New York: Palgrave Macmillan.

Hobson, J.M. (2004) *The Eastern Origins of Western Civilisation*, Cambridge: Cambridge University Press.

——(2007a) ‘Is Critical Theory Always for the White West and for Western Imperialism? Beyond Westphilian Towards a Post-racist Critical IR’, *Review of International Studies* 33(2): 91–116.

——(2007b) ‘Reconstructing International Relations Through World History: Oriental Globalization and the Global-Dialogic Conception of Inter-Civilizational Relations’, *International Politics* 44: 414–30.

——(2009) ‘Provincializing Westphalia: The Eastern Origins of Sovereignty’, *International Politics* 46: 671–90.

——(2012) *The Eurocentric Conception of World Politics: Western International Theory, 1760–2010*, Cambridge: Cambridge University Press.

Hodgson, M.G.S. (1993) *Rethinking World History Essays on Europe and World History*, Cambridge: Cambridge University Press.

Hogendorn, J. and Johnson, M. (1986) *The Shell Money of the Slave Trade*, Cambridge: Cambridge University Press.

Hopkins, A.G. (1973) *An Economic History of West Africa*, London: Longman.

Horniker, A.L. (1942) ‘William Harborne and the Beginning of Anglo-Turkish Diplomatic and Commercial Relations’, *The Journal of Modern History* 14(3): 289–316.

Howard, M. (1984) ‘The Military Factor in European Expansion’, in H. Bull and A. Watson (eds) *The Expansion of International Society*, Oxford: Clarendon Press.

Howard-Hassmann, R. (2008) *Reparations to Africa*, Philadelphia: University of Pennsylvania Press.

Howell, D.L. (1998) ‘Territoriality and Collective Identity in Tokugawa Japan’, *Daedalus* 127(3): 105–32.

Hsu, Immanuel (1960) *China’s Entry into the Family of Nations: The Diplomatic Phase, 1858–1880*, Cambridge, MA: Harvard University Press.

——(1990) *The Rise of Modern China*, 4th edn, Oxford: Oxford University Press.

——(1999) *The Rise of Modern China*, 6th edn, New York: Oxford University Press.

Huang, Jianhong (2003) ‘Zaoqi zhongguo zhengfu dui aomen de guanzhi yu aomen tongzhi de sheli’, *Wenhua zazhi* 49: 55–68.

Huang, S. (1987) ‘Kangxi’s Attitude in the Rites Controversy’, *Heythrop Journal* 28, 1: 57–67.

Hudson, Geoffrey F. (1965) *Europe and China: A Survey of their Relations from the Earliest Times to 1800*, London: Edward Arnold.

Hunt, George T. (1940) *The Wars of the Iroquois: A Study in Inter-Tribal Relations*, Madison: University of Wisconsin Press.

Huntington, S.P. (1991) ‘America’s Changing Strategic Interests’, *Survival* 33(1): 3–17.

——(1993) ‘Why International Primacy Matters’, *International Security* 17(4): 68–83.

——(1996) *The Clash of Civilizations and the Remaking of World Order*, New York: Simon and Schuster.

Ikenberry, G.J. (2008) ‘The Rise of China and the Future of the West: Can the Liberal System Survive?’ *Foreign Affairs* 87(1): 23–36.

Imber, C. (2002) *The Ottoman Empire, 1300–1650: The Structure of Power*, Basingstoke: Palgrave.

İnalcık, H. (1994) *An Economic and Social History of the Ottoman Empire*, Vol. 1: 1300–1600, Cambridge: Cambridge University Press.

——(2009) *Devlet-i'Aliyye: Osmanlı İmparatorluğu üzerine Araştırmalar*, Istanbul: Türkiye İş Bankası Kültür Yayınları.

——(2010) *Osmanlılar: Fütühat, İmparatorluk, Avrupa İle İlişkiler*, İstanbul: Timaş Yayınları.

Inikori, J. (1977) 'The Import of Firearms into West Africa 1750–1807', *Journal of African History* 18: 339–68.

——(2002) *Africans and the Industrial Revolution in England: A Study in International Trade and Economic Development*, Cambridge: Cambridge University Press.

——(2003) 'The Struggle Against the Slave Trade: The Role of the State', in S.A. Diouf (ed.) *Fighting the Slave Trade: West African Strategies*, Athens: Ohio University Press.

Irigoin, A. and Grafe, R. (2008) 'Bargaining for Absolutism: A Spanish Path to Nation-State and Empire Building', *Hispanic American Historical Review* 88(2): 173–209.

Jackson, P. (1999) 'From *Ulus* to Khanate: the Making of the Mongol States, c. 1220–c. 1290', in R. Amitai-Preiss and D. Morgan (eds) *The Mongol Empire and its Legacy*, Leiden: Brill.

Jackson, R. (1990) *Quasi-States: Sovereignty, International Relations and the Third World*, Cambridge: Cambridge University Press.

Jacques, M. (2009) *When China Rules the World: The End of the Western World and the Birth of a New Global Order*, London: Allen Lane.

Jahn, B. (1998) 'One Step Forward, Two Steps Back: Critical Theory as the Latest Edition of Liberal Idealism', *Millennium* 27(3): 613–41.

James, A. (1993) 'System or Society?' *Review of International Studies* 19(3): 269–88.

Jennings, F. (1975) *Invasion of America: Indians, Colonialism, and the Cant of Conquest*, Chapel Hill: University of North Carolina Press.

Jennings, F., Fenton, W.N., Druke, M.A. and Miller, D.R. (eds) (1985) *The History and Culture of Iroquois Diplomacy*, Syracuse, NY: Syracuse University Press.

Johnson, M. (1966) 'The Ounce in Eighteenth-Century West African Trade', *Journal of African History* 7: 197–214.

Jones, C.A. (2007a) 'Hierarchy and Resistance in American State-Systems, 1400–1800 CE', in S.J. Kaufman, R. Little and W.C. Wohlforth (eds) *The Balance of Power in World History*, Basingstoke: Palgrave Macmillan.

——(2007b) *American Civilization*, London: Institute for the Study of the Americas.

Jones, E.L. (1981) *The European Miracle: Environments, Economies and Geopolitics in the History of Europe and Asia*, Cambridge: Cambridge University Press.

Jones, K.L. (1994) 'Comparative Ethnohistory and the Southern Cone', *Latin American Research Review* 29(1): 107–18.

——(1999) 'Warfare, Reorganization and Readaptation at the Margins of Spanish Rule: The Southern Margin, 1573–1882', in F. Salomon and S.B. Schwartz (eds) *The Cambridge History of the Native Peoples of the Americas* (Volume 3: South America), Cambridge: Cambridge University Press.

Kaba, L. (2001) 'The Atlantic Slave Trade Was Not a "Black-on-Black Holocaust"', *African Studies Review* 44(1): 1–20.

Kacowicz, A.M. (1998) *Zones of Peace in the Third World: South America and West Africa in Comparative Perspective*, Notre Dame, IN: University of Notre Dame Press.

——(2005) *The Impact of Norms in International Society: The Latin American Experience, 1881–2001*, Notre Dame, IN: University of Notre Dame Press.

Kafadar, C. (1995) *Between Two Worlds: The Construction of the Ottoman State*, Berkeley: University of California Press

Kagan, R. (2008) *The Return of History and the End of Dreams*, New York: Alfred A. Knopf.

Kang, D.C. (2007) *The Rise of China*, New York: Columbia University Press.

Kappeler, A. (2001) *The Russian Empire: A Multiethnic History*, London: Longman.

Karasch, M. (2005) 'Rethinking the Conquest of Goiás, 1775–1819', *The Americas* 61(3): 463–92.

Karpat, K.K. (1973) *An Inquiry into the Social Foundation of Nationalism in the Ottoman State: From Social Estates to Classes, from Millets to Nation*, Princeton, NJ: Princeton University Press.

——(2001) *The Politicization of Islam: Reconstructing Identity, State, Faith, and Community in the Late Ottoman State*, Oxford: Oxford University Press.

Katō, E. (1981) 'Sakoku to bakuhansei kokka', in E. Katō and T. Yamada (eds) *Sakoku*, Tokyo: Yūhikaku.

Katzenstein, P.J. (2006) 'Multiple Modernities as Limits to Secular Europeanization', in T.A. Byrnes and P.J. Katzenstein (eds) *Religion in an Expanding Europe*, Cambridge: Cambridge University Press.

Kaufman, S.J., Little, R. and Wohlforth, W.C. (2007) *The Balance of Power in World History*, Basingstoke: Palgrave Macmillan.

Kayaoglu, T. (2010) 'Westphalian Eurocentrism in International Relations Theory', *International Studies Review* 12(2): 193–217.

Keal, P. (1995) '"Just Backward Children": International Law and the Conquest of Non-European Peoples', *Australian Journal of International Affairs* 49(2): 191–206.

——(2000) 'An International Society?' in G. Fry and J. O'Hagan (eds) *Contending Images of World Politics*, London: Macmillan.

——(2003) *European Conquest and the Rights of Indigenous Peoples: The Moral Backwardness of International Society*, Cambridge: Cambridge University Press.

Keene, D. (1969) *The Japanese Discovery of Europe, 1720–1830*, Stanford, CA: Stanford University Press.

Keene, E. (2002) *Beyond the Anarchical Society: Grotius, Colonialism and Order in World Politics*, Cambridge: Cambridge University Press.

Kennedy, P. (2010) 'Asia's Rise: Rise and Fall', *The World Today* 66(8/9): 6–9.

Klein, M. (2001) 'The Slave Trade and Decentralized Societies', *Journal of African History* 41(1): 49–65.

Klekar, C. (2006) '"Prisoners in Silken Bonds": Obligation, Trade, and Diplomacy in English Voyages to Japan and China', *Journal for Early Modern Cultural Studies* 6 (1): 84–105.

Kops, H.R.-De Bruyn (2002) 'Not Such an "Unpromising Beginning": The First Dutch Trade Embassy to China, 1655–1657', *Modern Asian Studies* 36(3): 535–78.

Krasner, S.D. (1995–96) 'Compromising Westphalia', *International Security* 20(3): 115–51.

——(1999) *Sovereignty: Organized Hypocrisy*, Princeton, NJ: Princeton University Press.

Kriger, C. (2006) *Cloth in African History*, Lanham, MD: Altamira Press.

Kunt, M. (1995a) 'Siyasal Tarih (1300–1600)', in S. Aksin (ed.) *Türkiye Tarihi, Vol. 2: Osmanli Devleti, 1300–1600,* Istanbul: Cem Yayinevi.

——(1995b) 'Siyasal Tarih (1600–1789)', in S. Aksin (ed.) *Türkiye Tarihi, Vol. 3: Osmanli Devleti, 1600–1908*, Istanbul: Cem Yayinevi.
Kupchan, C. (2012) *No One's World: The West, the Rising Rest and the Coming Global Turn*, New York: Oxford University Press.
Kurat, A.N. (1990) *Türkiye ve Rusya*, Ankara: Sevinç Matbaası.
Kürkçüoğlu, Ö. (2004) 'The Adoption and Use of Permanent Diplomacy', in A.N. Yurdusev (ed.) *Ottoman Diplomacy: Conventional or Unconventional?* New York: Palgrave Macmillan, 131–50.
Kurlantzick, J. (2007) *Charm Offensive: How China's Soft Power is Transforming the World*, New Haven, CT: Yale University Press.
Lach, D.F. (1965) *Asia in the Making of Europe: The Century of Discovery*, Volume I, Book 1 and Book 2, Chicago, IL: University of Chicago Press.
Lamar, H. and Truett, S. (1996) 'The Great Southwest and California from the Beginnings of European Settlement to the 1880s', in B.G. Trigger and W.E. Washburn (eds) *The Cambridge History of the Native Peoples of the Americas*, Vol. 1, Part 2: North America, Cambridge: Cambridge University Press.
Langfur, H. (1999) 'Myths of Pacification: Brazilian Frontier Settlement and the Subjugation of the Bororo Indians', *Journal of Social History* 32(4): 879–905.
——(2005) 'The Return of the Bandeira: Economic Calamity, Historical Memory, and Armed Expeditions to the Sertão in Minas Gerais, Brazil, 1750–1808', *The Americas* 61(3); *Rethinking Bandeirismo in Colonial Brazil* (January): 429–61.
Larkins, J. (2010) *From Hierarchy to Anarchy: Territory and Politics Before Westphalia*, Basingstoke: Palgrave Macmillan.
Larrain, J. (2006) 'Changes in Chilean Identity: Thirty Years after the Military Coup', *Nations and Nationalism* 12(2): 321–38.
Latham, A.J.H. (1973) *Old Calabar, 1600–1891*, Oxford: Oxford University Press.
Law, R. (1980) *The Horse in West African History*, Oxford: Oxford University Press.
——(1991) *The Slave Coast of West Africa, 1550–1750*, Oxford: Clarendon Press.
——(1994) '"Here is No Resisting the Country": The Realities of Power in Afro-European Relations on the West African Slave Coast', *Itinerario* 18: 50–64.
——(2004) *Ouidah: The Social History of a West African Slaving 'Port', 1727–1892*, Athens: Ohio University Press.
Lawson, P. (1993) *The East India Company: A History*, London and New York: Longman.
Layne, C. (1997) 'From Preponderance to Offshore Balancing: America's Future Grand Strategy', *International Security* 22(1): 86–124.
——(2012) 'The End of Pax Americana: How Western Decline Became Inevitable', *The Atlantic*, www.theatlantic.com/international/archive/2012/04/the-end-of-pax-americana-how-western-decline-became-inevitable/256388/ (accessed 30 January 2013).
Lenczowski, G. (1990) *The Middle East in World Affairs*, Ithaca, NY: Cornell University Press.
Lenman, B. and Lawson, P. (1983) 'Robert Clive, the "Black Jagir", and British Politics', *The Historical Journal* 26(4): 801–29.
Liang, C. (1959) *Intellectual Trends in the Ch'ing Period, 1644–1912*, trans. Immanuel C. Y. Hsu, Cambridge, MA: Harvard University Press.
Linklater, A. (2009) 'Grand Narratives and International Relations', *Global Change, Peace and Security* 21(1): 3–17.
Linklater, A. and Suganami, H. (2006) *The English School of International Relations: A Contemporary Reassessment*, Cambridge: Cambridge University Press.

Little, K. (1965) 'The Political Function of Poro: Part I', *Africa: Journal of the International Africa Institute* 35: 349–65.

——(1966) 'The Political Function of Poro: Part II', *Africa: Journal of the International Africa Institute* 36: 62–72.

Little, R. (2008) 'The Expansion of the International Society in Heeren's Account of the European States-system', *Working Paper No. 07–08*, School of Sociology, Politics and International Studies, University of Bristol.

Liu, Xiaomeng (2002) 'The Western Missionaries and Macao during the Period of Emperor Kangxi', *Revista de Cultura* 3: 126–35.

Liu, Yu (2008) 'The Intricacies of Accommodation: The Proselytizing Strategy of Matteo Ricci', *Journal of World History* 19(4): 465–87.

Lockhart, J. (1992) *The Nahuas after the Conquest: A Social and Cultural History of the Indians of Central Mexico, Sixteenth through Eighteenth Centuries*, Stanford, CA: Stanford University Press.

Lombardi, C. and Lombardi, J.V. (with Stoner, K.L.) (1983) *Latin American History: A Teaching Atlas*, Madison: University of Wisconsin Press for the Confederation on Latin American History.

López-Alves, F. (2001) 'The Transatlantic Bridge: Mirrors, Charles Tilly, and State Formation in the River Plate', in M.A. Centeno and F. López-Alvez (eds) *The Other Mirror: Grand Theory Through the Lens of Latin America*, Princeton, NJ: Princeton University Press.

Lovejoy, P. (2000) *Transformations in Slavery: A History of Slavery in Africa*, Cambridge: Cambridge University Press.

Lovejoy, P. and Falola, T. (eds) (2003) *Pawnship, Slavery and Colonialism in Africa*, Trenton: Africa World Press.

Lovejoy, P. and Richardson, D. (1999) 'The Business of Slaving: Pawnship in Western Africa, c.1600–1810', *Journal of African History* 42: 25–50.

——(2004) '"This Horrid Hole": Royal Authority, Commerce, and Credit at Bonny, 1690–1840', *Journal of African History* 45: 363–92.

——(2007) 'African Agency and the Liverpool Slave Trade', in D. Richardson, S. Schwarz and A. Tibbles (eds) *Liverpool and Transatlantic Slavery*, Liverpool: Liverpool University Press.

Lydon, G. (2009) *On Trans-Saharan Trails: Islamic Law, Trade Networks and Cross-Cultural Exchange in Nineteenth-Century West Africa*, Cambridge: Cambridge University Press.

Lyotard, J.-F. (1984) *The Postmodern Condition: A Report on Knowledge*, trans. G. Nennington and B. Massumi, Manchester: Manchester University Press.

MacGaffey, W. (1994) 'Dialogues of the Deaf: Europeans on the Atlantic Coast of Africa', in S. Schwartz (ed.) *Implicit Understandings: Observing, Reporting, and Reflecting on the Encounters Between Europeans and Other Peoples in the Early Modern Era*, Cambridge: Cambridge University Press.

Macartney, G. (1962) *An Embassy to China: Being the Journal Kept by Lord Macartney During his Embassy to the Emperor Ch'ienlung, 1793–1794*, ed. J.L. Cranmer-Byng, London: Longmans.

Maddison, A. (2007a) *Chinese Economic Performance in the Long Run: 960–2030 AD*, Paris: OECD.

——(2007b) *Contours of the World Economy 1–2030 AD: Essays in Macro-Economic History*, Oxford: Oxford University Press.

Mahbubani, K. (2008) *The New Asian Hemisphere: The Irresistible Shift of Global Power to the East*, New York: PublicAffairs.

Mamdani, M. (1996) *Citizen and Subject: Contemporary Africa and the Legacy of Late Colonialism*, Princeton, NJ: Princeton University Press.

Mancall, M. (1984) *China at the Centre: 300 Years of Foreign Policy*, New York: The Free Press.

Mann, K. (2007) *Slavery and the Birth of an African City: Lagos, 1760–1900*, Bloomington: University of Indiana Press.

Mann, M. (2005) *The Sources of Social Power,* Vol. 1: *A History of Power From the Beginning to A.D. 1760*, Cambridge: Cambridge University Press.

Mansel, P. (2010) *Levant: Splendor and Catastrophe on the Mediterranean*, London: John Murray Press.

Mardin, Ş. (1969) 'Power, Civil Society and Culture in the Ottoman Empire', *Comparative Studies in Society and History* 11(3): 258–81.

Mares, D.R. (2001) *Violent Peace: Militarised Interstate Bargaining in Latin America*, New York: Columbia University Press.

Marshall, P.J. (1968) *Problems of Empire: Britain and India, 1757–1813*, Vol. 2, London: Routledge.

——(1976) *East Indian Fortunes: The British in Bengal in the Eighteenth Century*, Oxford: Clarendon Press.

——(1998) 'British Expansion in India in the Eighteenth Century: A Historical Revision', in P.J.N. Tuck (ed.) *The East India Company, 1600–1858*, New York: Routledge.

Martin, P. (1972) *The External Trade of the Loango Coast, 1576–1870*, Oxford: Clarendon Press.

Massarella, D. (2001) '"Ticklish Points": The English East India Company and Japan, 1621', *Journal of the Royal Asiatic Society* 11(1): 43–50.

Mbembe, A. (2001) *On the Postcolony*, Berkeley: University of California Press.

McDowell, E. (1990) 'After the Cold War, the Land of the Rising Threat', *The New York Times*, 18 June.

McGowan, B. (1994) 'The Age of the Ayans, 1699–1812', in S. Faroqhi *et al.*, *An Economic and Social History of the Ottoman Empire*, Vol. 2: 1600–1914, Cambridge: Cambridge University Press.

McNeill, W.H. (1963) *The Rise of the West: A History of Human Community*, Chicago, IL and London: The University of Chicago Press.

——(1964) *Europe's Steppe Frontier 1500–1800*, Chicago, IL: University of Chicago Press.

——(1990) '*The Rise of the West* After 25 Years', *Journal of World History* 1(1): 1–21.

——(1991) 'The Rise of the West After 25 Years', in W.H. McNeill, *The Rise of the West: A History of the Human Community*, 2nd edn, Chicago, IL: Chicago University Press.

——(1993) 'Foreword', in A.G. Frank and B.K. Gills, *The World System: Five Hundred Years or Five Thousand*, London: Routledge

——(1998) 'The Changing Shape of World History', in P. Pomper, R.H. Elphick and R.T. Vann, *World History: Ideologies, Structures and Identity*, Oxford: Blackwell.

Merrell, J.H. (1989) 'Some Thoughts on Colonial Historians and American Indians', *The William and Mary Quarterly* 46(1): 94–119.

Meyerdorff, J. (1981) *Byzantium and the Rise of Russia: A Study of Byzantine-Russian Relations in the Fourteenth Century*, New York: Cambridge University Press.

Miers, S. and Kopytoff, I. (1977) 'African "Slavery" as an Institution of Marginality', in S. Miers and I. Kopytoff (eds) *Slavery in Africa: Historical and Anthropological Perspectives*, Madison: University of Wisconsin Press.

Millar, A.E. (2007) *The Jesuits as Knowledge Brokers Between Europe and China (1582–1773): Shaping European Views of the Middle Kingdom*, Working Papers No. 105/07. London: Department of Economic History, London School of Economics.
Miller, J. (1976) *Kings and Kinsmen: Early Mbundu States in Angola*, Oxford: Clarendon Press.
——(1988) *Way of Death: Merchant Capitalism and the Angolan Slave Trade, 1730–1830*, Madison: University of Wisconsin Press.
——(2004) 'Slavery and Colonialism', *Journal of Colonialism and World History* 5(3), muse.jhu.edu/login?auth=0&type=summary&url=/journals/journal_of_colonialism_and_colonial_history/v005/5.3intro.html (accessed 2 February 2013).
——(2012) *The Problem of Slavery as History: A Global Approach*, New Haven: Yale University Press.
Minamiki, G. (1985) *The Chinese Rites Controversy from Its Beginning to Modern Times*, Chicago, IL: Loyola University Press.
Mitchell, R. and Forbes, N. (eds) (1970) *The Chronicle of Novgorod, 1016–1471*, Hattiesburg, MS: Academic International.
Morgan, D. (1986) *The Mongols*, Oxford: Basil Blackwell.
Morgan, P. (2009) 'Africa and the Atlantic, c. 1450–1820', in J. Greene and P. Morgan (eds) *Atlantic History: A Critical Appraisal*, Oxford: Oxford University Press.
Morris, I. (2010) *Why the West Rules – For Now: The Patterns of History, and What They Reveal About the Future*, London: Profile Books.
Moyo, D. (2009) *Dead Aid: Why Aid is Not Working and How There is a Better Way for Africa*, New York: Farrar, Straus & Giroux.
Mungello, David E. (1989) *Curious Land: Jesuit Accommodation and the Origins of Sinology*, Honolulu: University of Hawaii Press.
——(1999) *The Great Encounter of China and the West: 1500–1800*, Lanham, MD: Rowman & Littlefield Publishers.
——(2005) *The Great Encounter of China and the West, 1500–1800*, 2nd edn, Lanham, MD: Rowman & Littlefield Publishers.
Naff, T. (1984) 'The Ottoman Empire and the European States System', in H. Bull and A. Watson (eds) *The Expansion of International Society*, Oxford: Clarendon Press.
Nagazumi, Y. (1990) *Kinsei shoki no gaikō*, Tokyo: Sōbunsha.
Naroll, R. (1969) 'The Causes of the Fourth Iroquois War', *Ethnohistory* 16(1): 51–81.
Nash, G.B. (1974) *Red, White, and Black: The Peoples of Early America*, Eaglewood Cliffs, NJ: Prentice Hall.
National Geographic (2004) *Historical Atlas of the United States*, Washington, DC.
Neumann, I.B. (1996) 'Self and Other in International Relations', *European Journal of International Relations* 2(2): 139–74.
——(1999) *Uses of the Other: The 'East' in European Identity Formation*, Minneapolis: University of Minnesota Press.
——(2008a) 'Russia's Standing as a Great Power, 1492–1815', in T. Hopf (ed.) *Russia's European Choices*, New York: Palgrave.
——(2008b) 'Russia as a Great Power, 1815–2000', *Journal of International Relations and Development* 11(3): 263–87.
——(2011) 'Entry in International Society Reconceptualised: The Case of Russia', *Review of International Studies* 37(2): 463–84.
Neumann, I.B. and Welsh, J.M. (1991) 'The Other in European Self-definition: An Addendum to the Literature on International Society', *Review of International Studies* 17: 327–48.

Neumann, I.B. and Wigen, E. (n.d., forthcoming) 'The Importance of the Eurasian Steppe to the Study of International Relations', *Journal of International Relations and Development.*
Nexon, D. and Wright, T. (2007) 'What's at Stake in the American Empire Debate?' *American Political Science Review* 101(2): 253–71.
Nkiwane, T. (2001) 'Africa and International Relations: Regional Lessons for a Global Discourse', *International Political Science Review* 22(2): 279–90.
Northrup, D. (1981) *Trade without Rulers: Pre-Colonial Economic Development in South-Eastern Nigeria*, Oxford: Clarendon Press.
——(2008) *Africa's Discovery of Europe, 1450-1850.* 2nd edn. Oxford: Oxford University Press.
Nunn, F.M. (1983) *Yesterday's Soldiers: European Military Professionalism in South America, 1890–1940*, Lincoln, NE, and London: University of Nebraska Press.
Nunn, N. (2007) 'Historical legacies: A Model Linking Africa's Past to its Current Underdevelopment', *Journal of Development Economics* 83: 157–75.
——(2010) 'Shackled to the Past: The Causes and Consequences of Africa's Slave Trade', in J. Diamond and J. Robinson (eds) *Natural Experiments of History*, Cambridge, MA: Harvard University Press.
Nwokeji, U. (2010) *The Slave Trade and Culture in the Bight of Biafra: An African Society in the Atlantic World*, New York: Cambridge University Press.
——(2011) 'Slavery in Non-Islamic West Africa, 1420–1820', in D. Eltis and S. Engerman (eds) *The Cambridge World History of Slavery: Volume 3*, Cambridge: Cambridge University Press.
O'Brien, P. (1984) 'Europe in the World Economy', in H. Bull and A. Watson (eds) *The Expansion of International Society*, Oxford: Clarendon Press.
O'Hagan, J. (1995) 'Civilisational Conflict? Looking for Cultural Enemies', *Third World Quarterly* 16(1): 19–38.
——(2000) 'A "Clash of Civilizations"?' in G. Fry and J. O'Hagan (eds) *Contending Images of World Politics*, London: Macmillan.
——(2002) *Conceptualizing the West in International Relations: From Spengler to Said*, Basingstoke: Palgrave.
——(2005) 'The Question of Culture', in A.J. Bellamy (ed.) *International Society and its Critics*, Oxford: Oxford University Press.
Omer-Cooper, John D. (1966) *The Zulu Aftermath: A Nineteenth-century Revolution in Bantu Africa*, London: Longman.
Ortaylı, İ. (1987) *İmparatorluğun En Uzun Yüzyılı*, İstanbul: Hil Yayınları.
——(2006) *Son İmparatorluk Osmanlı*, İstanbul: Timaş Yayınları.
——(2007) *Osmanlı Barışı*, İstanbul: Timaş Yayınları.
Osabu-Kle, D.T. (2000) 'The African Reparation Cry: Rationale, Estimate, Prospects, and Strategies', *Journal of Black Studies* 30(3): 331–50.
Osiander, A. (2001) 'Sovereignty, International Relations and the Westphalian Myth', *International Organization* 55(2): 251–87.
Ostrowski, D.G. (1998) *Muscovy and the Mongols: Cross-Cultural Influences on the Steppe Frontier, 1304–1589*, Cambridge: Cambridge University Press.
——(2000) 'Muscovite Adaptation of Steppe Political Institutions: A Reply to Halperin's Objections', *Kritika: Explorations in Russian and Eurasian History, New Series* 1(2): 267–97.
Padden, R.C. (1957) 'Cultural Change and Military Resistance in Araucanian Chile, 1550–1730', *Southwestern Journal of Anthropology* 13(1): 103–21.

Pamuk, O. (2000) *A Monetary History of the Ottoman Empire*, Cambridge: Cambridge University Press.

Patterson, O. (1982) *Slavery and Social Death: A Comparative Study*, Cambridge, MA: Harvard University Press.

Paul, H. (1977) 'De Coningh on Deshima: *Mijn Verblijf in Japan*, 1856', *Monumenta Nipponica* 32(3): 347–64.

Payton, J.R. (2006) 'Ottoman Millet, Religious Nationalism and Civil Society: Focus on Kosova', *Religion in Eastern Europe* 26(1): 1–13.

Pelenski, J. (1967) 'Muscovite Imperial Claims to the Kazan Khanate', *Slavic Review* 26(4): 559–76.

Peterson, W.J. (1998) 'Learning from Heaven: The Introduction of Christianity and Other Western Ideas into Late Ming China', in D. Twitchett and F.W. Mote (eds) *Cambridge History of China*, Vol. 8, The Ming Dynasty, 1368–1644, Part 2, Cambridge: Cambridge University Press.

Peyrefitte, A. (1993) *The Collision of Two Civilisations: The British Expedition to China 1792–4*, London: Harvill Press.

Philpott, D. (2001) *Revolutions in Sovereignty: How Ideas Shaped Modern International Relations*, Princeton, NJ: Princeton University Press.

Poe, M.T. (2000) *'A People Born to Slavery': Russia in Early European Ethnography, 1476–1748*, Ithaca, NY: Cornell University Press.

Polanyi, K. (1957) *The Great Transformation: The Political and Economic Origins of Our Time*, Boston, MA: Beacon Press.

Pomeranz, K. and Topik, S. (2006) *The World Created: Society, Culture and the World Economy 1400 to the Present*, New York: M.E. Sharpe.

Price, R. and Reus-Smit, C. (1998) 'Dangerous Liaisons? Critical International Theory and Constructivism', *European Journal of International Relations* 4(3): 259–94.

Quataert, D. (1994) 'The Age of Reforms, 1812–1914', in H. İnalcık and D. Quataert (eds) *An Economic and Social History of the Ottoman Empire,* Vol. 2*: 1600–1914*, Cambridge: Cambridge University Press.

——(2000) *The Ottoman Empire: 1700–1922*, Cambridge: Cambridge University Press.

Quirk, J. (2008) 'Historical Methods', in D. Snidal and C. Reus-Smit (eds) *Oxford Handbook of International Relations*, Oxford: Oxford University Press.

——(2009) *Unfinished Business: A Comparative Survey of Historical and Contemporary Slavery*, Paris: UNESCO Publishing.

——(2011) *The Anti-Slavery Project: From the Slave Trade to Human Trafficking*, Philadelphia: Pennsylvania University Press.

Quirk, J. and Richardson, D. (2009) 'Anti-Slavery, European Identity and International Society: A Macro-Historical Perspective', *The Journal of Modern European History* 7(1): 70–94.

Quirk, J. and Vigneswaran, D. (2013) 'Contemporary Bondage in Africa: Historical Legacies and Recent Innovations', in J. Quirk and D. Vigneswaran (eds) *Slavery, Migration and Contemporary Bondage in Africa*, Trenton: Africa World Press.

Rachman, G. (2011) 'American Decline: This Time is for Real', *Foreign Policy* 184: 59–72.

Radelet, S. (2010) 'Success Stories from "Emerging Africa"', *Journal of Democracy* 21 (4): 87–101.

Rahman, Nor-Afidah Abd (2003) 'The China Trade', infopedia.nl.sg/articles/SIP_377_2005-01-18.html (accessed 2 February 2013).

Reeve, M.-E. (1994) 'Regional Interaction in the Western Amazon: The Early Encounter and the Jesuit Years, 1538–1767', *Ethnohistory* 41(1): 106–38.

Reus-Smit, C. (1999) *The Moral Purpose of the State: Culture, Social Identity, and Institutional Rationality in International Relations*, Princeton, NJ: Princeton University Press.

——(2002) 'Imagining Society: Constructivism and the English School', *British Journal of Politics and International Relations* 4(3): 487–509.

——(2005) 'Liberal Hierarchy and the Licence to use Force', *Review of International Studies* 31: 71–92.

Ricci, M. (1985) *The True Meaning of the Lord of Heaven*, ed. D. Lancashire, Guozhen Hu and E. Malatesta, Ann Arbor, MI: Institute of Jesuit Sources, University of Michigan.

Rich, P. (1999) 'European Identity and the Myth of Islam: A Reassessment', *Review of International Studies* 25(3): 435–51.

Richardson, D. (1979) 'West African Consumption Patterns and their Influence on the Eighteenth-Century English Slave Trade', in H. Gemery and J. Hogendorn (eds) *The Uncommon Market: Essays in the Economic History of the Atlantic Slave Trade*, New York: Academic Press.

——(1991) 'Slave Prices in West and West-Central Africa, 1700–1810: Towards an Annual Series', *Bulletin of Economic Research* 43(1): 21–56.

Richter, D.K. (1983) 'War and Culture: The Iroquois Experience', *The William and Mary Quarterly* 40(4): 528–59.

Risse, T., Ropp, S.C. and Sikkink, K. (1999) *The Power of Human Rights: International Norms and Domestic Change*, Cambridge: Cambridge University Press.

Roberston, C. and Klein, M. (1988) *Women and Slavery in Africa*, Wisconsin: University of Wisconsin Press.

Rock, D. and López-Alves, F. (2000) 'State-Building and Political Systems in Nineteenth-Century Argentina and Uruguay', *Past and Present* 167(1): 176–202.

Rodney, W. (1970) *History of the Upper Guinea Coast, 1545–1800*, Oxford: Oxford University Press.

——(1972) *How Europe Underdeveloped Africa*, London: Bogle-L'Ouverture.

Roniger, L., Green, J.N. and Yankelevich, P. (eds) (2012) *Exile and the Politics of Exclusion in the Americas*, San Diego, CA: Sussex Academic.

Rowbotham, A.H. (1942) *Missionary and Mandarin: The Jesuits at the Court of China*, Berkeley: University of California Press.

Rubinstein, A.Z. (1982) *Soviet Policy Toward Turkey, Iran, and Afghanistan: The Dynamics of Influence*, New York: Prager Publishers.

Ruggie, J.G. (1993) 'Territoriality and Beyond: Problematizing Modernity in International Relations', *International Organization* 47(1): 139–74.

Saeger, J.S. (1999) 'Warfare, Reorganization and Readaptation at the Margins of Spanish Rule: The Chaco and Paraguay, 1573–1882', in F. Salomon and S.B. Schwartz (eds) *The Cambridge History of the Native Peoples of the Americas*, Vol. 3: South America, Cambridge: Cambridge University Press.

Said, E.W. (1995) *Orientalism: Western Conceptions of the Orient*, London: Penguin.

Salisbury, N. (1996) 'Native People and European Settlers in Eastern North America', in B.G. Trigger and W.E. Washburn (eds) *The Cambridge History of the Native Peoples of the Americas*, Vol. 1: North America, Part 1, Cambridge: Cambridge University Press.

Salomon, F. and Schwartz, S.B. (eds) (1999) *The Cambridge History of the Native Peoples of the Americas*, Vol. 3: South America, Cambridge: Cambridge University Press.

Schlesier, K.H. (1975) 'The Indians of the United States: An Essay on Cultural Resistance', *Wichita State University Bulletin*, University Studies No. 81.

——(1976) 'Epidemics and Indian Middlemen: Rethinking the Wars of the Iroquois, 1609–53', *Ethnohistory* 23(2): 129–45.

Schmidt, V.H. (2006) 'Multiple Modernities or Varieties of Modernity?' *Current Sociology* 54(1): 77–97.

Schouenborg, L. (2013) *The Scandinavian International Society: Primary Institutions and Binding Forces, 1815–2010*, Abingdon: Routledge.

Searing, J. (1993) *West African Slavery and Atlantic Commerce: The Senegal River Valley, 1700–1860*, Cambridge: Cambridge University Press.

Sebes, J. (ed.) (1961) *The Jesuits and the Sino-Russian Treaty of Nerchinsk (1689): The Diary of Thomas Pereira*, Rome: Institutum Historicum S.I.

Sen, S. (1998) *Empire of Free Trade: The East India Company and Making of the Colonial Marketplace*, Philadelphia: University of Pennsylvania Press.

Sen, S.P. (1962) 'The Role of Indian Textiles in Southeast Asian Trade in the Seventeenth Century', *Journal of Southeast Asian History* 3(2): 92–110.

Shaw, S.J. (1976) *History of the Ottoman Empire and Modern Turkey*, Vol. 1, *Empire of the Gazis: The Rise and Decline of the Ottoman Empire, 1280–1808*, Cambridge: Cambridge University Press.

Shu, L. (1994) 'Ferdinand Verbiest and the Casting of Cannons in the Qing Dynasty', in J.W. Witek (ed.) *Ferdinand Verbiest (1623–1688): Jesuit Missionary, Scientist, Engineer, and Diplomat*, Nettetal: Steyler Verlag.

Sidbury, J. (2007) *Becoming African in America: Race and Nation in the Early Black Atlantic*, Oxford: Oxford University Press.

Simmons, D. (1956) 'An Ethnographic Sketch of the Efik People', in C.D. Forde (ed.) *Efik Traders of Old Calabar*, London: Oxford University Press.

Skilliter, S. (1977) *William Harborne and the Trade with Turkey, 1578–1582: A Documentary Study of the First Anglo-Ottoman Relations*, Oxford: Oxford University Press.

Slave Voyages (n.d.) www.slavevoyages.org (accessed 28 May 2012).

Slotkin, R. (1973) *Regeneration Through Violence: The Mythology of the American Frontier, 1600–1860*, Middletown, CT: Wesleyan University Press.

——(1985) *The Fatal Environment: The Myth of the Frontier in the Age of Industrialization, 1800–1890*, New York: Atheneum.

Smallwood, S. (2007) *Saltwater Slavery: A Middle Passage from Africa to American Diaspora*, Cambridge, MA: Harvard University Press.

Smith, K. (2009) 'Has Africa got Anything to Say? African Contributions to the Theoretical Development of International Relations', *The Round Table* 98(402): 269–84.

Snajder, M. and Roniger, L. (2009) *The Politics of Exile in Latin America*, Cambridge: Cambridge University Press.

Sommerdyk, S. and da Silva, F.R. (2010) 'Reexamining the Geography and Merchants of the West Central African Slave Trade: Looking Behind the Numbers', *African Economic History* 38: 77–105.

Souza, G.B. (1986) *The Survival of Empire – Portuguese Trade and Society in China and the South China Sea 1630–1754*, Cambridge: Cambridge University Press.

Spence, J.D. (1980) *To Change China: Western Advisers in China*, London: Penguin.

——(1999) *The Chan's Great Continent: China in Western Minds*, New York and London: W.W. Norton & Co.

Spruyt, H. (1994) *The Sovereign State and its Competitors: An Analysis of Systems Change*, Princeton, NJ: Princeton University Press.

Spuler, B. (1965a) *Die Goldene Horde: Die Mongolen in Russland 1223–1502*, two vols, Wiesbaden: Otto Harrassowitz.

——(1965b [1943]) *Die Goldene Horde. Die Mongolen in Russland 1223–1502*, second enlarged edn, Wiesbaden: Otto Harrassowitz.

Stannard, D.E. (1987) 'The Invisible People of Early American History', *American Quarterly* 39(4): 649–55.

Steele, I.K. (1994) *Warpaths: Invasions of North America*, Oxford: Oxford University Press.

Subramanian, L. (1987) 'Banias and the British – The Role of Indigenous Credit in the Process of Imperial Expansion in Western India in the 2nd-Half of the 18th-Century', *Modern Asian Studies* 21(3): 473–510.

——(2001) 'Arms and the Merchant: The Making of the Bania Raj in Late Eighteenth-Century India', *South Asia-Journal of South Asian Studies* 24(2): 1–27.

Suganami, H. (1984) 'Japan's Entry in International Society', in H. Bull and A. Watson (eds) *The Expansion of International Society*, Oxford: Clarendon Press.

——(2003) 'British Institutionalists, or the English School, 20 Years On', *International Relations* 17(3): 253–71.

Sutherland, L. (1952) *The East India Company in Eighteenth-Century Politics*, Oxford: Clarendon Press.

Suzuki, S. (2005) 'Japan's Entry into Janus-faced European International Society', *European Journal of International Relations* 11(1): 137–64.

——(2009) *Civilization and Empire: China and Japan's Encounter with European International Society*, London: Routledge.

——(2012) 'Europe and Japan', *Review of European Studies* 4(3): 54–63.

Sweet, J. (2003) *Recreating Africa: Culture, Kinship, and Religion in the African-Portuguese Word, 1441–1770*, Chapel Hill: University of North Carolina.

Tanaka, T. (1975) *Chūsei taigai kankeishi*, Tokyo: Tōkyō daigaku shuppankai.

Tang, Kaijian (1999) *Aomen kaibu chuqishi yanjiu*, Beijing: Zhonghua shuju.

Tashiro, K. (1982) 'Foreign Relations During the Edo Period: Sakoku Reexamined', trans. S.D. Videen, *Journal of Japanese Studies* 8(2): 283–306.

Taylor, A.C. (1999) 'The Western Margins of Amazonia from the Early Sixteenth to the Early Nineteenth Century', in F. Salomon and S.B. Schwartz (eds) *The Cambridge History of the Native Peoples of the Americas*, Vol. 3: South America, Cambridge: Cambridge University Press.

Taylor, I. (2010) *The International Relations of Sub-Saharan Africa*, New York: Continuum.

Taylor, I. and Williams, P. (2004) *Africa and International Politics: External Involvement on the Continent*, London: Routledge.

Teng, Ssu-yu and Fairbank, John King (eds) (1954) *China's Response to the West: A Documentary Survey, 1839–1923*, Cambridge, MA: Harvard University Press.

Teschke, B. (2003) *The Myth of 1648: Class, Geopolitics and the Making of Modern International Relations*, London, New York: Verso.

Therborn, G. (2003) 'Entangled Modernities', *European Journal of Social Theory* 6(3): 293–305.

Thomson, J.E. (1994) *Mercenaries, Pirates and Sovereigns: State-Building and Extra-Territorial Violence in Early Modern Europe*, Princeton, NJ: Princeton University Press.

Thornton, J. (1998) *Africa and Africans in the Making of the Atlantic World, 1400–1800*, Cambridge: Cambridge University Press.

——(1999) *Warfare in Atlantic Africa: 1500–1800*, London: Routledge.

——(2007) 'The Portuguese in Africa', F. Bethencourt and D.R. Curto (eds) *Portuguese Oceanic Expansion, 1400–1800*, Cambridge: Cambridge University Press.

Toby, R.P. (1984) *State and Diplomacy in Early Modern Japan: Asia in the Development of the Tokugawa Bakufu*, Stanford, CA: Stanford University Press.

——(1991) *State and Diplomacy in Early Modern Japan: Asia in the Development of the Tokugawa Bakufu*, Stanford, CA: Stanford University Press.

Tompson, R.S. (1986) *The Atlantic Archipelago: A Political History of the British Isles*, Lewiston, New York: E. Mellen.

Treadgold, Donald W. (1973) *The West in Russia and China: Religious and Security Thought in Modern Times, Volume 2: China, 1582–1949*, Cambridge: Cambridge University Press.

Trevor-Roper, H. (1965) *The Rise of Christian Europe*, New York: Harcourt, Brace & World.

Trigger, B.G. and Washburn, W.E. (eds) (1996) *The Cambridge History of the Native Peoples of the Americas*, Vol. 1, Part 1: North America, Cambridge: Cambridge University Press.

——(1996) *The Cambridge History of the Native Peoples of the Americas*, Vol. 1, Part 2: North America, Cambridge: Cambridge University Press.

Tsukamoto, M. (1979) 'Edo jidai ni okeru "i" kannen ni tsuite', *Nihon rekishi* 371: 1–18.

Turnbull, P. (1975) *Warren Hastings*, London: New English Library.

Uyar, M. and Erickson, E.J. (2009) *A Military History of the Ottomans: From Osman to Atatürk*, Oxford: Preager Security International Press.

van den Boogaart, E. (1992) 'The Trade between Western Africa and the Atlantic World, 1600–690: Estimates of Trends in Composition and Value', *The Journal of African History* 33(3): 369–85.

van Goor, J. (2004) 'Merchants as Diplomats: Embassies as Illustrations of European-Asian Relations', *Review of Culture* 11: 49–64.

Vansina, J. (1990) *Paths in the Rainforests: Towards a Political Tradition in Equatorial Africa*, London: James Currey.

Vernadsky, G. (1948) *Kievan Rus: A History of Russia, Vol. II*, New Haven, CT: Yale University Press.

——(1953) *The Mongols and Russia: A History of Russia, Vol. III*, New Haven, CT: Yale University Press.

Vigezzi, B. (2005) *The British Committee on the Theory of International Politics (1954–1985): The Rediscovery of History*, trans. Ian Harvey, Milan: Edizioni Unicopli.

Vincent, R.J. (1982) 'Race in International Relations', *International Affairs* 58(4): 658–70.

——(1984) 'Racial Equality', in H. Bull and A. Watson (eds) *The Expansion of International Society*, Oxford: Clarendon Press.

Voegelin, E. (1941) 'The Mongol Orders of Submission to European Powers, 1245–55', *Byzantion. Revue internationale des etudes Byzantines* 15 (1940–41): 378–413.

Wakeman, F. (1978) 'The Canton Trade and the Opium War', in John King Fairbank (ed.) *The Cambridge History of China*, Vol. 10, Late Ch'ing, 1800–1911, Part I, Cambridge: Cambridge University Press.

Waley-Cohen, J. (1999) *The Sextants of Beijing: Global Currents in Chinese History*, London: W.W. Norton & Co.

Wallerstein, I. (1974) *The Modern World-System: Capitalist Agriculture and the Origins of the European World-Economy in the Sixteenth Century*, New York: Academic Press.

Wan, Ming (2001) *Zhong pu zaoqi guanxi shi*, Beijing: Shehui Kexue Wenxian Chubanshe.

Wang, D. (2005) *China's Unequal Treaties: Narrating National History*, Lanham, MD: Lexington Books.

Wang, Dongqing and Pan, Rudan (2004) 'Mingchao haijin zhengce yu jindai xifang guojia de diyici duihua junshi chongtu', *Gudai Junshi* 2: 138–52.

Watson, A. (1964) *The War of the Goldsmith's Daughter*, London: Chatto and Windus.

——(1984a) 'European International Society and its Expansion', in H. Bull and A. Watson (eds) *The Expansion of International Society*, Oxford: Clarendon Press.

——(1984b) 'Russia and the European States System', in H. Bull and A. Watson (eds) *The Expansion of International Society*, Oxford: Clarendon Press.

——(1984c) 'New States in the Americas', in H. Bull and A. Watson (eds) *The Expansion of International Society*, Oxford: Clarendon Press.

——(1992) *The Evolution of International Society: A Comparative Historical Analysis*, London: Routledge.

——(2009) *The Evolution of International Society: A Comparative Historical Analysis*, London: Routledge (new edn of 1992).

Watson, I.B. (1980) 'Fortifications and the "Idea" of Force in Early East India Company Relations with India', *Past and Present* 88(1): 70–87.

Wendt, A. (1999) *Social Theory of International Politics*, Cambridge: Cambridge University Press.

Wight, M. (1966) 'Western Values in International Relations', in H. Butterfield and M. Wight (eds) *Diplomatic Investigations: Essays in the Theory of International Politics*, London: Allen and Unwin.

——(1977) *System of States*, ed. H. Bull, Leicester: Leicester University Press.

——(1992) *International Theory: The Three Traditions*, New York: Holmes & Meier.

Wills, J.E. (1968) 'Ch'ing Relations with the Dutch, 1662–90', in J.K. Fairbank (ed.) *The Chinese World Order: Traditional China's Foreign Relations*, Cambridge, MA: Harvard University Press.

——(1979) 'Maritime China from Wang Chih to Shih Lang: Themes in Peripheral History', in J. Spence and J.E. Wills (eds) *From Ming to Ch'ing: Conquest, Region, and Continuity in Seventeenth Century China*, New Haven, CT and London: Yale University Press.

——(1984) *Embassies and Illusions: Dutch and Portuguese Envoys to K'ang-hsi, 1666–1687*, Cambridge, MA: Council on East Asian Studies, Harvard University.

——(1998) 'Relations with Maritime Europeans, 1514–1662', in D. Twitchett and F.W. Mote (eds) *Cambridge History of China*, Vol. 8, The Ming Dynasty, 1368–1644, Part 2, Cambridge: Cambridge University Press.

——(2011) *China and Maritime Europe, 1500–1800: Trade, Settlement, Diplomacy, and Missions*, New York and Cambridge: Cambridge University Press.

Witek, J.W. (1996) 'Johann Adam Schall von Bell and the Transition from the Ming to the Ch'ing Dynasty', in R. Malek (ed.) *Western Learning and Christianity in China: The Contribution and Impact of Johann Adam Schall von Bell, S.J. (1592–1666)*, Nettetal: Steyler Verlag.

Wohlforth, W.C., Little, R., Kaufman, S.J., King, D., Jones, C.A., Tin-Bor Hui, V., Eckstein, A., Deudney, D. and Brenner, W.J. (2007a) 'Testing Balance-of-Power Theory in World History', *European Journal of International Relations* 13(2): 155–85.

Wohlforth, W., Little, R. and Kaufmann, S. (eds) (2007b) *The Balance of Power in World History*, London: Palgrave.

Wright, R.M. (1999) 'Destruction, Resistance, and Transformation: Southern, Central, and Northern Brazil, 1580–1890', in F. Salomon and S.B. Schwartz (eds) *The Cambridge History of the Native Peoples of the Americas*, Vol. 3: South America, Cambridge: Cambridge University Press.

Young, C. (1994) *The African Colonial State in Comparative Perspective*, New Haven, CT: Yale University Press.

Young, J.D. (1980) *East–West Synthesis: Matteo Ricci and Confucianism*, Hong Kong: Centre for Asian Studies, University of Hong Kong.

——(1983) *Confucianism and Christianity: The First Encounter*, Hong Kong: Hong Kong University Press.

Yurdusev, A.N. (2003) *International Relations and the Philosophy of History*, Houndmills: Palgrave.

——(2004) 'The Ottoman Attitude Toward Diplomacy', in A.N. Yurdusev (ed.) *Ottoman Diplomacy: Conventional or Unconventional?* New York: Palgrave Macmillan, 5–35.

——(2009) 'The Middle East Encounter with the Expansion of European International Society', in B. Buzan and A. Gonzalez-Pelaez (eds) *International Society and the Middle East: English School Theory at the Regional Level*, Basingstoke: Palgrave Macmillan.

Zakaria, F. (2009) *The Post-American World*, New York: W.W. Norton & Co.

Zeitlin, J.F. (1989) 'Ranchers and Indians on the Southern Isthmus of Tehuantepec: Economic Change and Indigenous Survival in Colonial Mexico', *Hispanic American Historical Review* 69(1): 23–60.

Zhang, T. (2004) '"Puren ouzhu aomen haidao de ao shuo": yanjiu silu de lishi yanbian', *Wenhua zazhi* 51: 79–86.

——(2005) 'On the Portuguese Trade in China, 1513–20', *Revista de Cultura* 15: 180–88.

Zhang, X. (2009) *Ouzhou zaoqi hanxue shi: zhongxi wenhua jiaoliu yu xifang hanxue de xingqi*, Beijing: Zhonghua shuju.

Zhang, Y. (1991) *China in the International System, 1918–1920: The Kingdom at the Periphery*, Basingstoke: Macmillan.

——(2001) 'System, Empire and State in Chinese International Relations', *Review of International Studies* 27(5): 43–63.

Zhang, Y. and Buzan, B. (2012) 'The Tributary System as International Society in Theory and Practice', *Chinese Journal of International Politics* 5(1): 3–36.

Zhou, Y. (ed.) (1987) *Zhongwai wenhua jiaoliu shi*, Zhengzhou: Henan Remin Chubanshe.

Zhou, Z. (ed.) (2010) *Kuayue kongjian de wenhua: 16–19 shiji zhongxi wenhua de xiangyu yu tiaoshi*, Shanghai: Dongfang chuban zhongxin.

Zorin, V.A. *et al.* (eds) (1959) *Istoriya diplomatii*, second edn, Moscow: Gospolitizdat.

Index